200 TROPICAL PLANTS

200 CONSPICUOUS, UNUSUAL, OR ECONOMICALLY IMPORTANT

TROPICAL PLANTS

OF THE CARIBBEAN

Illustrated with Photographs
Reproduced in Full Color

by

John M. Kingsbury

Professor of Botany, Emeritus
Cornell University

BULLBRIER PRESS
Ten Snyder Heights, Ithaca, NY 14850
1988

Nearly all the photographs in this book were exposed on Kodachrome 64 film, usually with ambient light. All the color work in its production, including digital separation of photographs and control of inks in the printing process, was performed electronically by computer. The manuscript was composed on a word processor and the typesetting was done directly from the resulting computer disks.

Typesetting, color separations, printing, and binding are the work of Glundal Color, 6700 Joy Road, East Syracuse, NY.

This book is available singly or in quantity from Bullbrier Press, Ten Snyder Heights, Ithaca, New York 14850 USA.

Printed and bound in the United States of America.

ISBN 0-9612610-2-1

200 TROPICAL PLANTS

To Louise G. Kingsbury, whose hand is visible, literally, in several of the photographs and, figuratively, in all other aspects of the accomplishment of this book.

Other books by John M. Kingsbury:

Deadly Harvest
Poisonous Plants of the United States and Canada
Seaweeds of Cape Cod and the Islands
The Rocky Shore

INTRODUCTION

This book has grown from a need to satisfy the curiosity of participants in adult education programs of Cornell University that have taken place in various locations in the American tropics over the past decade. Its contents thus reflect the actual questions of several hundred travellers about the plants they notice.

The text consists largely of descriptions of plants. These have been written in plain English as much as possible. Even so, each one is botanically accurate and detailed enough to allow the user to decide in nearly every instance whether a plant-in-hand is or is not the plant described.

This work is organized like an encyclopedia with the plants listed by scientific name in strict alphabetical order. This eliminates the need to hunt for pages by number. Thus, the pages are not numbered.

Most people prefer to use common names instead of scientific names. Unfortunately, common names of plants are notoriously variable and unreliable. This is especially true of plants, like most of those described here, that are distributed world-wide in the tropics in lands populated by people of diverse ethnic backgrounds and languages. Each of these plants usually has several or many common names depending on where you are and with whom you are talking. Common names also change with the times and circumstances. A good example of a new name is the recently rechristened "tourist tree".

Scientific names, in contrast, have been stabilized by international agreement and are legally controlled throughout the world. A given plant species will have exactly the same name in Russia as in the U.S., in Tasmania as in Iceland, or in Tortola as in Martinique. A scientific name consists uniformly of two words. The first, always with the initial letter capitalized, is the name of the genus (pl. genera), analogous to a human patronym. The second is the species name (pl. the same), analogous to a person's first name. Closely related genera are gathered into discrete groups, termed the family. Plant family names usually (but not always) end in "aceae".

The standard reference used to establish the appropriate scientific name for each plant described here (wherever possible) is: *Hortus III,* by the Staff of the Liberty Hyde Bailey Hortorium, Cornell University, published by Macmillan, New York.

HOW TO USE THIS BOOK

To start from scratch, (if you don't know or cannot find out even a common name), page through the illustrations for an approximate match. Then compare your plant with the description in the text. If it agrees, fine. If not, keep looking. The text often suggests look-alikes to check out and gives ways of distinguishing among them. If you still cannot make a match, perhaps your plant is not among those included here.

To look up a plant if you know the common name, use the *Index to Common Names* (green pages) to find the corresponding scientific name and then go directly to it in the alphabetical page headings. Usually you need remember no more than the first four letters of the first name (genus name) to do this successfully. Even if you do not find a listing for the common name you know, the plant may still be in this book. Some common names are so localized or uncertain in use that they have not been included in the text or index. Check out the pictures for a match.

A few of the indexed common names go with plants that don't have their own separate headings in the text. If you find no page heading for the scientific name you found in the index to common names, look for it next in the *Index to Synonyms and Locations*. You will then be directed to the correct page heading under which to find it. This index will also help in finding plants listed under older or variant scientific names in other books. If you do not find the scientific name of a plant you are seeking either in the page headings or in this index, you may safely conclude that the plant is not mentioned or described in this book.

If you are curious about the relatives of any plant in which you are interested, find the name of its plant family in its description. Then turn to the third index, the *Index to Relationships by Family*. Here the plant families are listed alphabetically, with the names of all the other plants in each family that are also included in this book.

SPECIAL TERMS

A few special terms, not easy to replace accurately with ordinary English words, have been used in the text.

Leaf structure:

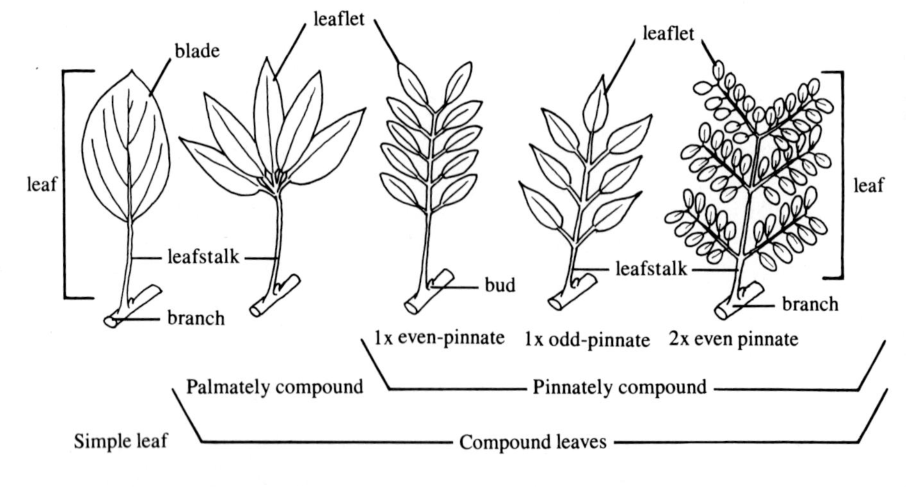

Leaf drop:

A plant that sheds all its leaves simultaneously, leaving bare branches, is "deciduous". One that drops its leaves one by one, always leaving many behind, is "evergreen". It is not correct to equate "evergreen" with "conifer" as many people from cold latitudes do.

Leaf blades:

The term "elliptical", to describe the shape of the leaf or leaflet blade, is used loosely. In most plants, the exact shape of the blade varies considerably within the species, or even in a single individual plant. Other words that might have been used in particular instances include oblong, oval, narrow, wide, etc.

"Heart-shaped" (of a blade) implies an evident incutting dent or notch where the leaf stalk is attached to the blade.

A blade is "simple" when, although it may be lobed or incut, it is not divided all the way to the leafstalk into separated units ("leaflets").

A blade edge may be scalloped, wavy, toothed, lobed, etc. An edge not indented in any of these ways is described as "smooth".

Flower:

A discrete cluster of several to many individual flowers is an "inflorescence".

A "bract" is a modified leaf, usually colored other than green, associated with the flower or inflorescence (see Anthurium andraeanum, Alpinia purpurata, or Euphorbia pulcherrima).

A basic understanding of the structure of a flower is presumed. Words like "sepal", "petal", "calyx", "corolla" "stamen", "anther", and "style", have been used sparingly but when necessary. Definitions and illustrations explaining these terms can be found in any good elementary high school or college text on general biology.

Fruit:

"Pod" is a general word for a fruit that may be elongate or globose, straight or twisted, flat or fleshy, large or small, succulent or dry, large or tiny.

Tuber:

"Tuber" has a precise botanical definition but the word is used in a looser, common English sense here. It may include structures more precisely called corms, for example.

Species:

When a genus name is followed by "sp.", as in "Araucaria sp.", it signifies that the plant is a species belonging to that genus, but its exact identity is not determined at the species level. When a genus name is followed by "spp.", as in "Araucaria spp.", it means that the text refers to two or more species of Araucaria collectively, without listing them separately by name.

ACKNOWLEDGMENTS

My thanks go first to all those acute participants in Cornell's Adult University Programs who asked questions about tropical plants and thus shaped this book.

They go second to all those unnamed and unknown persons who developed and maintained botanical gardens in tropical locations. I have profited especially from repeated visits to St. George Village Botanical Garden on St. Croix, Andromeda Garden on Barbados, and the Botanic Garden of Tortola.

They go third to those persons who made resources at the College of the Virgin Islands on St. Croix freely available.

Individuals who helped especially with advice and encouragement include Nicholas Clarke, Laurence Hodge, Walter Knausenberger, Jennie Lane, Darshan Padda, and Arthur Petersen.

The author took all the photographs but one (of Agave missionum by Nicholas Clarke) and is responsible for any errors that may remain in the text. He would appreciate learning of them so they may be corrected.

PLANT DESCRIPTIONS

Abrus precatorius

Rosary pea
Crab's eyes
Precatory bean

FAMILY: Leguminosae, Faboideae

HABIT: Weedy vine, to 15 ft.

HABITAT: Widespread throughout the tropics; native and naturalized.

NATIVE TO: Tropical Asia.

LEAF: Alternate, 1x even-pinnate, about 3 in. long; each with 10 to 30 pairs of blunt, almost square-tipped leaflets, each less than 1 in. long, paper-thin.

FLOWER: Elongate clusters of tiny red, pink, lavender, or white pea-like flowers, each about 1/2 in. long.

FRUIT: Flat, beaked pods, each about 1 1/2 in. long; the pods twist and split open when dry exposing 4 to 6 conspicuous black and red seeds.

STEM: Green, graying in older parts; often twining about others.

SEEDS: Very regular in size and shape, about 1/4 in. wide, glossy as though enameled, jet black at attachment and over about one-third of the seed; the rest bright red.

•The glossy seeds are extremely attractive and are commonly used in necklaces, bracelets, and other craft work. Unfortunately, they are also very toxic! One seed if thoroughly masticated can kill a child. The bright red changes to a red-brown in old necklaces over time. Articles made with these dangerous seeds may be confiscated by customs officials. A few related legumes produce seeds of similar color, but usually larger. The plant commonly twines among other vegetation and is inconspicuous except when in seed.

Acacia tortuosa

Thornbush
Casha
Sweet briar

FAMILY: Leguminosae, Mimosoideae

HABIT: Shrub or small tree, to 15 ft.

HABITAT: Dry woodlands; savanna.

NATIVE TO: Caribbean.

LEAF: Alternate, 2x even-pinnate; leaflets numerous, each about 1/4 in. long and blunt tipped, borne on about 5 to 6 pairs of pinnae; commonly winter-deciduous.

FLOWER: Cylindrical pendant inflorescence of tiny yellow flowers; the whole about 5 in. long.

FRUIT: Nearly cylindrical legume pod, 4 to 5 in. long, slightly narrowed between seeds.

STEM: Woody, contorted, sometimes zig-zag, mostly horizontally disposed; with copious whitish woody thorns an inch or more long on mature stems.

•This is a common weedy invader of grasslands. It is protected against grazing by its densely placed fierce thorns. In older parts of the plant, mature branches commonly die or fail to produce leaves when new branches grow forth elsewhere on the shrub. Healthy plants thus sometimes look half dead. This is an adaptation to the dry conditions characteristic of its habitat. Other species of Acacia include the common "wattles" of Australia, many of which inhabit similar dry terrain.

Acalypha hispida

Chenille plant
Monkey's tail
Red hot cattail

FAMILY: Euphorbiaceae

HABIT: Shrub, to 15 ft.

HABITAT: Ornamental.

NATIVE TO: Indonesia.

LEAF: Simple, alternate, green, to 8 in. long. The blade is pointed, broadly oval or heart-shaped, with toothed edges and pinnately arranged veins; borne on a short but distinct leaf stalk. The leafstalk and veins are hairy.

FLOWER: Sexes on separate plants. Male flowers inconspicuous. Female flowers crowded in large showy hanging purple-red catkins (white in one variety), about 1 1/2 ft. long and 1 in. in diameter. As in many members of this family, the petals are absent and sepals are small and insignificant. Each flower is subtended by a small bract and the female parts (style and ovary) are conspicuous and hairy.

FRUIT: Absent or inconspicuous.

•This ornamental shrub may be grown singly, or used as a hedge. Male plants are not found in cultivation. Female plants are propagated by cuttings. The name comes from the Greek word for nettles which its leaves resemble. When not in flower, it might be confused with an Hibiscus. The toothed leaves are similar in size and shape, but the long, showy, pendant female inflorescence is like no other. In the related Acalypha wilkesiana (next), the female inflorescence is shorter, less showy, and more hidden within the leaves.

Acalypha wilkesiana

Joseph's coat
Copper leaf
Match-me-if-you-can
Beefsteak

FAMILY: Euphorbiaceae

HABIT: Shrub, to 15 ft.

HABITAT: Ornamental.

NATIVE TO: Pacific islands.

LEAF: Simple, alternate, oval or narrow-elongate, long- or short-stalked; blades with scalloped, toothed, or nearly smooth edges; 5 to 8 in. long or more in narrow kinds. Varieties have leaves splotched or margined with green, copper, white, red, purple, or yellow in various mixes.

FLOWER: Each plant has separate male and female flower clusters. Male flowers are inconspicuous in erect clusters; females small, clustered in hanging reddish spikes 1/4 inch wide by 8 inches long, partly hidden in the leaves.

FRUIT: Inconspicuous, three-chambered.

•This species has a score or more of horticultural varieties under several names, varying tremendously in leaf color, shape, patterns of marking, and length of stalk. In some varieties the leaf is short-stalked, heart-shaped, thickish-textured, and somewhat twisted. The leaves are often densely packed, especially near the ends of the branches. In the commonest variety, the leaf is long-stalked, conspicuously pointed, thin-textured, flat (not twisted), and brightly colored. A variety with splotchy white leaves can be easily confused with the variegated variety of Hibiscus rosa-sinensis when not in flower. Another confusion is possible with a variety of Codiaeum or with Graptophyllum, but the leaves of both of these are never toothed.

Adansonia digitata

Baobab tree
Monkey bread tree
Dead rat tree
Guinea tamarind

FAMILY: Bombacaceae

HABIT: Massive tree, to 60 ft.

HABITAT: Individuals planted for native medicinal and religious significance; also naturalized.

NATIVE TO: Africa.

LEAF: Palmately compound with 5 to 7 glossy leaflets, each about 5 in. long, attached closely to a long leaf stalk; commonly deciduous in the dry season.

FLOWER: Large white flowers appear after leaf drop, hanging on long stringy stems; petals to 6 in. across, spreading or bent back; with a spherical cluster of stamens on an elongated stalk at the center.

FRUIT: An unusual velvety cylinder, about 4 by 16 in. hanging from long "strings" in the branches. Each has a rounded collar where the stem attaches.

TRUNK: Massive, squat, to as much as 30 ft. thick, often soon branched into several main trunks.

•This is one of Africa's most useful trees. The trunk becomes hollow with age and it may store as much as 250 gallons of water thus supporting much wildlife. The trees are slow-growing and become monumental in size and appearance given time. Baobab has been planted sporadically throughout the tropics as a curiosity or object of religious veneration. The persistent gray or khaki-colored fuzzy fruits hanging in the branches look, at a distance, like dead rats hung up by their tails. They can be confused with the fruits of Kigelia which hang in a similar way.

Agave americana
Agave missionum

Century plant
Aloe
Maypole

FAMILY: Agavaceae

HABIT: Herbaceous perennial.

HABITAT: Ornamental; dry scrublands.

NATIVE TO: The Caribbean and Central and South America.

LEAF: Large, fleshy, strap-like, clustered at the ground; each to 5 ft. or more long by 10 in. wide; dull green or grayish, curved erect, or or bent back (especially when flowering); edged with spines; tip sharply spine-pointed.

FLOWER: The main stem elongates rapidly 20 ft. or more at flowering; flowers clustered in lateral groups that look like upturned candelabras; each flower 6 parted, tubular, pale yellow.

FRUIT: Bulbils (incipient plantlets) are commonly formed as sprouts in the inflorescence of A. americana.

•Despite its common name, this is not the genus Aloe (which see). There are some 300 species of Agave, difficult to separate. The common large wild species in the Virgin Islands is A. missionum. The common landscape species is A. americana. Other common ornamental species tend to be smaller than these. Most species display marginal prickles, but not all. The leaves are white-edged in some.

Until flowering, each plant has a single stem bud which produces leaves in a cluster near the ground. After one or more decades (but not really a century), when enough foodstuffs have been manufactured by and stored in the leaves, the stem bud suddenly starts growing rapidly, producing the inflorescence stem. When that is done, it dies and the whole plant dies with it, although a new plant may sprout from the roots. Sisal fibers and the liquors pulque and mescal come from other species of Agave. Agave can be confused with Yucca and some of the smaller species, commonly potted, can be confused with Aloe, bromelliads, and cacti.

Albizia lebbeck

Tibet
Mother-in-law's tongue
Shakshak

FAMILY: Leguminosae, Mimosoideae

HABIT: Tree, to 100 ft.

HABITAT: Ornamental; shade tree.

NATIVE TO: Tropical Asia and Australia.

LEAF: Alternate, 2x even-pinnate; to 15 in. long; with 4 to 9 pairs of oblong leaflets, each about 2 in. long. The foliage is deciduous in winter.

FLOWER: Flowers clustered in spherical greenish-yellow "pincushions" consisting mostly of radiating stamens, each nearly 2 in. in diameter. Petals tiny, white.

FRUIT: Flat legume pods to 1 ft. long by about 2 in. wide; usually numerous; green, then yellow, turning light brown; seeds appearing "shrink-packed" within the pods.

•The tree is usually broad-topped and spreading in aspect. The light-colored pods persist on the trees a long time (even after leaf drop) and usually are a conspicuous feature of this species. When dry, they rattle constantly in the breezes giving rise to the second and third common names above. The dark heartwood is prized for various uses. This is a common shade tree throughout the tropics.

Allamanda cathartica

Allamanda
Yellow allamanda
Golden trumpet

FAMILY: Apocynaceae

HABIT: Shrub or vine; 10 to 15 ft.

HABITAT: Ornamental.

NATIVE TO: Brazil.

LEAF: Simple, opposite, or in whorls; glossy, to 5 in. long; shiny green, elongate, pinnately veined, tapering at both ends with almost no leaf stalk; not deciduous.

FLOWER: Usually scattered, single, large, bright yellow, trumpet-shaped, arising from a small 5-pointed green cup-like calyx; single or double; 3 in. across by 4 in long; with 5 rounded waxy thickish petal lobes overlapping like a pinwheel at the base in single forms.

FRUIT: Globose, spiny, splitting open in two; seeds flat, brown, with thin yellow margins.

•Allamanda flowers occur in clusters, but only one is usually open at a time in each cluster. The flower buds are pointed, glossy, and reddish brown. In most varieties this color persists on the outside of the petals and sometimes inside the flower tube as well. There are many horticultural varieties of this species, including one with doubled flowers (see illustration). As implied in its species name, in its native habitat this plant is used as a cathartic.

Allamanda violacea is a purple Allamanda, but is not the plant usually meant by that name (Cryptostegia grandiflora, which see).

Alocasia macrorrhiza

Colocasia esculenta (syn. C. antiquorum)

Elephant's ear
Taro
Dasheen
Poi
Tannia
Ape

FAMILY: Araceae

HABIT: Herbaceous perennial, 4 to 15 ft.

HABITAT: Cultivated both as an ornamental and for food.

NATIVE TO: Southeast Asia and Pacific islands.

LEAF: Simple, arising in a cluster at the ground or at the top of a short, squat, trunk; long-stalked, erect or spreading out and down, to 3 ft. or more; blade large, heart- or spear-shaped, usually with three points, one at the tip and two backward; the latter sometimes rounded instead. The main veins go to the three points.

FLOWER: Typical spathe-spadix inflorescence of this family (see Anthurium); yellowish or yellow-green in both species.

FRUIT: Inconspicuous.

STEM: Thick, fleshy, vertical, mostly or entirely buried; marked with elongate horizontal leaf scars. Youngest leaves arise from the top.

•These two genera, Alocasia and Colocasia, are closely related, and each contains several other species which are usually smaller than the two described here. Each of the two described species has several distinct varieties that are grown as ornamentals and others for food.

Ornamental varieties of either of these species (usually called elephant's ear) vary in the way the leaf blades are displayed, and their coloring and pattern. In some, the blades are bluish with veins of contrasting color. Others are variegated with lighter or darker patches (and may be confused with Caladium which is a close relative).

The most important agricultural varieties (usually of Colocasia) are grown for the tuber, commonly called taro, which yields poi and similar foodstuffs. The majority of these varieties are cultivated in paddies that are flooded from time to time. Others are adapted to upland conditions.

Taro reached the Nile Valley from Southeast Asia by 500 BC. It was distributed throughout the Pacific islands by native migrations and is a major source of starch for human consumption in that region. Any part of either species (including raw taro) contains a sap that imparts an unpleasant odor and causes an intensely painful reaction in the mouth if eaten raw. The cooked, pounded tuber yields poi, a starchy paste. The tuber can also be used directly if boiled, baked, or roasted. Leaves can be cooked like cabbage.

Taro is harvested 1 to 1½ yr. after planting. The top is cut off the tuber, trimmed, and replanted to start the next crop.

The botanical difference between Colocasia and Alocasia has to do with technical details of the ovary. Another less consistent but more useful character relates to the attachment of the leaf blade on its stalk. In a mature leaf of Colocasia, the leaf stalk is attached to the blade at a short distance in from the edge where the lobes join (top left illustration). In Alocasia, the blade is attached at its edge between the backward pointing lobes (middle left photograph). The bottom left photograph shows some naturalized Colocasia on Guadaloupe.

Aloe barbadense

Aloe
Barbados aloe
Unguentine cactus
Medicinal aloe

FAMILY: Liliaceae

HABIT: Herbaceous perennial.

HABITAT: Ornamental; commercial.

NATIVE TO: Mediterranean.

LEAF: Rosette of erect or spreading grayish green, often waxy, conspicuously fleshy, strap-like leaves, 1 to 3 ft. long; not fibrous; narrowing to a fine point; edges with spiny white or reddish teeth; the surface may be splotched with white. The sap is jelly-like.

FLOWER: A single (or few-branched) erect cluster of dense flowers, to 30 in. tall; each flower yellow, narrow-tubular, lily-like, about 1 in. long.

•This species forms clusters of leaves at ground level from a horizontal underground stem and reproduces mainly by offsets. Other species of Aloe (of which there are many) may be stemmed above ground. Compare Aloe with Agave and Yucca which are superficially similar in foliage.

The copious transparent jelly-like substance in the leaves (see illustration) is a common constituent of cosmetics and aloe is also used in drugs. The plant is grown commercially primarily in the Netherlands Antilles. It is a major element in native medicines of the tropics, used in many ways for many purposes. The plant is often kept close by in pots at doorways. It is also used as an ornamental around the bases of buildings or as a landscape planting. Although spelled the same, the scientific name is pronounced with three syllables, the common name with two.

Alpinia purpurata

Red ginger
Ginger lily

FAMILY: Zingiberaceae

HABIT: Herbaceous perennial, to 15 ft.

HABITAT: Ornamental.

NATIVE TO: Pacific islands.

LEAF: Simple, alternate, elliptical; blade about 2½ ft. long by 6 in. broad, with almost no leaf stalk. The blade is smooth-surfaced, pinnately veined.

FLOWER: A large red brush, showy, erect, (eventually hanging), to 1 ft. long. Individual flowers small white, tubular, short-lived, mostly hidden inside heavy persistent red bracts that are each over 1 in. long and constitute the main floral feature.

FRUIT: New plantlets may form in the older parts of the inflorescence.

•Nicolai elatior (torch ginger) is similar, but makes an erect cone of more tightly appressed bracts.

True ginger is a species of Zingiber to which Alpinia is closely related. The root of Alpinia purpurata has a distinct odor of ginger and looks like ginger root as well. The conspicuous pinnate venation of the leaves separates Alpinia from most lilies and amaryllids which have parallel venation. (But so do some other members of the ginger family.)

The white flowers (see closeup) are delicate and evanescent, and often fully hidden within the red bracts. Bracts last a long time. This means that most inflorescences have no actual flowers showing most of the time.

Alpinia zerumbet (syn. A. speciosa)

Shell ginger
Pink porcelain lily

FAMILY: Zingiberaceae

HABIT: Herbaceous perennial, to 12 ft.

HABITAT: Ornamental.

NATIVE TO: East Asia.

LEAF: Simple, alternate; blade lance-shaped, shiny, about 5 in. wide by 2 ft. long, pinnately veined; almost no leaf stalk.

FLOWER: Inflorescence horizontal or hanging; bracts small and insignificant but the flowers themselves are conspicuous; sepals tubular, white tipped with purple, lasting; petal tube yellow or orange, extending a little beyond the sepals; short-lived.

FRUIT: Small, inconspicuous.

•The unusual porcelain-like color and texture of the sepals distinguish this common ornamental from a number of others of similar foliage with which it might otherwise easily be confused. The gingers usually grow as clumps of cane-like stems bearing large leaf blades. Left unpruned, this ginger forms a dense rank weedy growth. Characteristics useful in separating ornamental gingers that are not flowering from similar plants are described under Alpinia purpurata.

Ananas comosus (syn. A. sativus)

Pineapple
Pine

FAMILY: Bromeliaceae

HABIT: Perennial herb to 4 ft.

HABITAT: Cultivated.

NATIVE TO: Southern Brazil.

LEAF: Simple, coming from the stem in a tight spiral, thus forming a cluster almost at ground level; each blade without stalk, long and slender, harsh with small spines along the edges, very sharp-pointed, about 3 ft. long by 1 1/2 in. wide.

FLOWER: Small, tubular, inconspicuous, short-lived, violet or reddish; maturing from base of fruit upwards. Inflorescence bracts often conspicuously colored red or purple.

FRUIT: Compound (from many flowers united), to 1 ft. tall, erect, topped with a headdress of leaves.

•A number of distinct named varieties of pineapple have been developed by the plant breeder for commercial purposes. Simultaneous flowering, needed for mechanized harvesting, is induced by spraying the fields with a synthetic hormone. Each plant produces a single terminal fruit. The edible flesh represents the product of about 100 compacted flowers. Commercial varieties are not effectively pollinated (requires hummingbirds) and therefore are seedless. Plants are propagated instead by cuttings from the fruit tops or from side shoots. Commercial production requires plastic mulch and irrigation. Flowering is induced at about 1 yr.; the first harvest is at 20 months, and the second harvest about a year later from a side shoot plant. Then the field is replowed. Some varieties yield useful fiber from the leaves.

Columbus first discovered pineapples on his second voyage to the Caribbean (1493).

Annona spp.

Sour sop
Sweet sop
Sugar apple
Custard apple
Bullock's heart
Cherimoya

FAMILY: Annonaceae

HABIT: Small trees.

HABITAT: Cultivated.

NATIVE TO: Tropical America.

LEAF: Alternate, simple, somewhat two-ranked on the stem, elliptical, smooth edged, 4 to 10 in. long depending on species, smooth (blades velvety beneath in A. cherimoya); usually persistent (not deciduous) except in A. squamosa;

FLOWER: Single, 3-parted, about 1 in. long, yellow or greenish-yellow in all four species.

FRUIT: Large, ovoid or spherical, conspicuous, compound (composed of the joined product of many flowers; see below); edible. All have small hard black seeds, one from each original flower. Fruits are green when young, turning greenish yellow or yellowish when ripe, in all four species.

•The fruits of Annona trees can be confused with those of Morinda (pain killer) and Artocarpus (breadfruit). Here are some differences that will serve to separate these genera readily: The leaves of pain killer and breadfruit are proportionately larger and wider than those of any Annona, and in breadfruit, the fruits also are distinctly larger. Also, pain killer leaves are opposite, and those of breadfruit are deeply cut in a pinnate fashion.

Four species of Annona are commonly cultivated in the tropics for the fruit: 1. A. muricata – sour sop; fruit ovoid or elongate, often somewhat asymmetric, to 8 in.long, dark green, covered with curved fleshy spines (which may have been rubbed off by the time the fruits appear in the markets); used mainly for juice (illustrated top right and left). 2. A. squamosa – sweet sop, sugar or custard apple; fruit globose or heart-shaped, to 5 in. long, no spines; units well marked and partly separating when ripe (illustrated in the three photographs in the lower half page). 3. A. reticulata – bullock's heart or custard apple; fruit heart-shaped to ovoid, 5 in. across, surface smooth, but with distinct unit and spine scar markings; flesh custard-like (illustrated middle right). 4. A. cherimola, – cherimoya, fruit like the latter (not illustrated).

Soursop is the commonest of the Annona species in most places. The juice of its fruits has a sweet-sour taste and is prized for refreshing drinks and for flavoring sherbets. The flesh is somewhat fibrous. The flesh of Annona squamosa (sweet sop) is sugary and custard-like. This one is used primarily as a dessert fruit. Also used primarily as a dessert fruit is that of Annona cherimola (cherimoya) which is very sweet and juicy but sometimes somewhat granular in texture. It is considered the best flavored of all. Unfortunately, Annona fruits are also prized by livestock, which can present a problem if animals get loose in an orchard.

Anthurium andraeanum

Anthurium
Flamingo flower
Oilcloth flower

FAMILY: Araceae

HABIT: Herbaceous perennial, to 2 ft.

HABITAT: Cultivated ornamental.

NATIVE TO: Colombia.

LEAF: Clustered, from ground level; long-stalked; blades hanging, leathery, elongate heart-shaped, evergreen, to 10 in. long by half as wide; tip pointed, with rounded lobes.

FLOWER: Consisting of a single showy spreading heart-shaped spathe (a modified leaf), 3 to 5 in. long, glossy polished, puckered between the veining. The finger-like, straight or slightly curved, golden and ivory spadix, which arises from it, is 2¼ in. long. In a closely related species (A. scherzerianum) the spadix is twisted like a pig-tail. See more below.

FRUIT: Inconspicuous.

•The structure of the unusual inflorescence of Anthurium is characteristic of the family (Araceae), and consists of a cigar-like spike of densely packed minute flowers (the spadix) arising from a single, specially modified leaf (the spathe) which serves the necessary attractive function for all the numerous, almost insignificant-looking individual flowers.

There are numerous hybrids and cultivars of Anthuriums, including salmon, white, or pink kinds. The ancestral color, is red – still the commonest and perhaps the brightest or most intense of any. The almost unreal inflorescences (which look as though made of oilcloth) will last three weeks or longer after cutting, and can be shipped by air to markets around the world from the point of origin.

Spathiphyllum (which see) is similar in appearance, but the spathe is pure white.

Antigonon leptopus

Coral vine
Mexican creeper
Mexican love chain
Coralita

FAMILY: Polygonaceae

HABIT: Weedy vine, 20 to 40 ft.

HABITAT: Ornamental; naturalized.

NATIVE TO: Mexico.

LEAF: Simple, alternate, stalked, smooth-edged, heart- or arrow-shaped, 4 in. long; veins conspicuously depressed in the surface of the leaf blade; the blade edges somewhat wavy.

FLOWER: Hanging festoons of tiny bright rosy-pink flowers each consisting of a calyx of 5 heart-shaped sepals (no petals). Most flowers are usually closed. When the sepals are spread back, they reveal 8 yellow-tipped stamens. The inflorescence stem is often continued on as a tendril at its tip, an unusual feature (see illustration).

FRUIT: Small three-angled seed, contained in an inflated papery envelope.

STEM: Slim, much-branched, climbing by tendrils.

•This species is occasionally planted, but may be encountered more frequently as a weed entangled in the vegetation along roadsides and similar places where it brightens the scene. There are light pink and white flowered kinds as well. Antigonon possesses scores of common names in various parts of the tropics. The tubers are edible and, like its relative the buckwheat, its flowers are very attractive to bees.

Araucaria heterophylla
Araucaria columnaris (syn. A. excelsa)

Norfolk Island pine (A. heterophylla)
Cook's pine (A. columnaris)

FAMILY: Araucariaceae

HABIT: Large trees, to 200 ft. tall.

HABITAT: Ornamental.

NATIVE TO: Norfolk Island and New Caledonia, respectively.

LEAF: Dense, needle-like, incurved, lying all around and along the stem and overlapping more or less; about 1/4 in. long in A. columnaris and 1/2 in. in A. heterophylla. These are juvenile leaves; the mature leaves (rarely seen) are flattened, and nearly as wide as long.

FLOWER: Plants with separate sexes; male cone is nearly 2 in. long; the female subglobose, 3 to 5 in. long.

FRUIT: Araucarias can be told from pines by the number of seeds per cone scale; one here, while pine has two underneath each scale.

STEM: The main branches come out in a regular and distinctive whorled pattern when trees are not crowded or severely shaded from one side. The trunk can become massive with age, to a diameter of 9 ft.

•These two species are closely related and difficult to tell apart. Typical mature Cook's pine (named for Capt. James Cook who discovered it on New Caledonia) is broadly columnar (illustration top left); Norfolk Island pine is distinctly pyramidal (top middle). Both are excellent timber trees. Araucaria araucana is the monkey puzzle tree of Chile. Its leaves are longer (nearly one inch) than either of these two species.

Artocarpus altilis

Breadfruit
Breadnut

FAMILY: Moraceae

HABIT: Tree, 50 to 60 ft.

HABITAT: Cultivated.

NATIVE TO: East Indies.

LEAF: Alternate, simple, large ovate but deeply pinnately lobed, short-stalked, to 3 ft. long; blade thick, leathery; upper surface glossy, under-surface rough.

FLOWER: Sexes separate on same tree; male inflorescence club-shaped, upright, 6 to 12 in. long, with thousands of small yellow-green flowers; female similar but spherical.

FRUIT: Round to oval, to 8 in. long and up to 10 lb. in weight; surface obviously spiny when young, but coarsely pebbled with tiny spines at maturity unless rubbed off in handling; yellow or brownish when fully ripe.

•Breadfruit is a selected seedless variety, propagated by root suckers; breadnut is a seeded variety, otherwise similar. Its seeds roasted are somewhat like chestnuts. The leaves in bud are covered by a conspicuous leaf sheath.

Capt. Bligh's illfated voyage on the *Bounty* was to bring breadfruit plants to the Caribbean to serve as food for the plantation slaves. It still is a starchy diet staple for many Pacific peoples. The wood is light and easily worked. Leaf sheaths can be used as fine sandpaper. An abundant white sap serves as canoe caulking, chewing gum, and bird-trapping glue. Ripeness determines best fruit use: fully ripe, the fruit is used similarly to sweet potato but must be peeled and cored before cooking.

Artocarpus heterophyllus (illustrated in the bottom middle and bottom right photographs) is the less common jack-fruit tree. The leaves are very much smaller, not lobed, and the fruits are commonly produced directly from the tree trunk or the largest branches. They are very large, oblong, up to 2 ft. in length and 40 lb. in weight. The fruits of both species of Artocarpus can be confused with those of Morinda and Annona (which see).

Asystasia gangetica

Sutter's gold

FAMILY: Acanthaceae

HABIT: Low-growing or creeping weedy herb to 2 ft.

HABITAT: Cultivated as a ground cover; naturalized.

NATIVE TO: India, Southeast Asia, Africa.

LEAF: Simple, opposite, membranous, smooth-edged, broadly elliptical to heart-shaped, short-stalked, 1 to 3 in. long.

FLOWER: One-sided, few-flowered clusters; petal tube flaring to 1 in. across, 5-parted, the lobes often slightly grouped into two upper and three lower, white, yellow, or lavender, fading to pale purple.

FRUIT: Elliptical pod, about 1 in. long, containing a few flattened round seeds.

STEM: Often terminating in tendril-like new growth.

•This plant is unusual in the number of flower colors to be seen in a single population at the same time. It is commonly grown as a ground cover, especially on bankings. It survives close clipping as a hedge and can even be mowed to within a few inches of the ground to make a lawn of sorts.

Averrhoa carambola

Carambola
Star fruit

FAMILY: Oxalidaceae

HABIT: Small tree, to 30 ft.

HABITAT: Cultivated.

NATIVE TO: Southeast Asia.

LEAF: Alternate, 1x odd-pinnate; 5 to 11 leaflets, blades elliptical, 1 to 2 in. long. The leaflets are short-stalked narrowing to a pointed tip. They often hang from either side of the leaf-stalk and look like individual leaves.

FLOWER: Single or in small clusters, tiny, 5-parted, variegated white and purple.

FRUIT: Ovoid, to 5 in. long with 3 to 5 deep ribs; yellow to yellow-brown when ripe, with seeds in a watery pulp.

•This plant is grown as a market crop in some areas. It grows best in shade. The flesh of the fruit is crisp and juicy. Its sourness is due to oxalic acid, characteristic of the family. The fruit appears in the markets occasionally, now increasingly, in northern cities. It is used as a juice, in preserves, or in salads where thin slices with their star-like outline have special eye appeal. There are sweet and sour cultivars. Unripe fruit or particularly sour ones may be used as pickles. Another species, A. bilimbi, (belimbi, blimbing, or cucumber tree by common name) is less star-shaped, distinctly more acid, and usually used only as a pickle.

Bambusa vulgaris

Bamboo
Common bamboo

FAMILY: Gramineae

HABIT: Perennial woody grass, forming branching stems to 70 ft. tall.

HABITAT: Cultivated; naturalized.

NATIVE TO: Unknown, but probably tropical and subtropical Asia.

LEAF: Simple, alternate, clustered somewhat at stem ends; blade narrowing sharply and attached almost directly to the stem; blade almost linear, tapering to a point, 10 by 1¼ in., with rough edges and under-surface; veins parallel.

FLOWER: Bamboo rarely flowers. The inflorescence is a large, open spray of clusters of grass spikelets.

STEM: Open clumps of greenish or yellow stems, often splotched with black and ringed by leaf scars. Stems are jointed, to 5 in. in diameter, straight when young, then bending with the weight of foliage. Branch buds are covered at first with a broad papery sheath which persists at its base as the branch grows forth.

•Bamboo is a giant grass, spreading by underground runners. Its young shoots grow upward through other vegetation and reach light, then spreading and leafing out. They will then form a dense weedy thicket if not kept in check. Many other species of large perennial tropical grasses are sometimes also called bamboos. Common bamboo is useful to native peoples in hundreds of ways such as for construction, for carrying water, and for making musical instruments. It was domesticated early and is no longer known in the wild.

Bauhinia variegata
Bauhinia purpurea

Poor man's orchid
Mountain ebony
Bull hoof
Orchid tree

FAMILY: Leguminosae, Caesalpinioideae

HABIT: Tree, to 40 ft.

HABITAT: Ornamental.

NATIVE TO: India, China.

LEAF: Simple, alternate, short-stalked; blades 4 to 6 in. across, shorter than broad. distinctly incut at the tip; in outline like a cow's footprint; halves often partly folded upward like a butterfly's wings at rest. The major veins radiate from the point of attachment.

FLOWER: In short clusters at branch tips; each flower 4 to 6 in. across; with 5 lavender to magenta petals; and with upcurving stamens and style projecting from the center. The top petal is usually distinctly different in color from the others.

FRUIT: Smooth, flat, green then brown pod, about 1 ft. by 1 in. which eventually snaps open, scattering shiny black seeds.

STEM: Bauhinia trees typically have a short trunk and spreading crown.

•Bauhinia includes many species and varieties of related trees and shrubs. The trees are usually difficult to identify at the species and variety levels. Of these two, the shape and size of the petals differ somewhat. The petals are about 1 in. wide in Bauhinia variegata, often touching or overlapping a bit; they are typically narrower in Bauhinia purpurea.

There are many other species of Bauhinia that differ from the above general description. Bauhinia punctata, for example, is a sprawling shrub, native to tropical Africa, with brick red flowers, about 2½ in. across. Bauhinia punctata is illustrated at the lower left. The other illustrations are of Bauhinia sp. One thing all Bauhinia species share is the unusual hoof-print leaf. The genus is named after John and Caspar Bauhin, 16th century botanists, reflecting the closeness (brotherly nature) of the two leaf lobes.

Bixa orellana

Annatto
Arnotto
Lipstick tree

FAMILY: Bixaceae

HABIT: Shrub or small tree, to 20 ft.

HABITAT: Cultivated and ornamental.

NATIVE TO: Tropical America.

LEAF: Simple, alternate, pinnately veined, 3 to 7 in. long; stalked; blade heart-shaped or broadly elliptical, pointed.

FLOWER: In open terminal clusters; each looks something like a large apple blossom, pink, rose, or white, about 2 in. across, 5-parted; sepals and petals slightly overlapping; numerous stamens within.

FRUIT: A pointed pod, bright red or red-brown, about 2 in. long, covered with soft spines, splitting in two when dry; containing 30 to 50 seeds each covered with a thin waxy red coating.

•Bixa is grown commercially for an intense yellow dye obtained from the seeds, and long used by native peoples for personal decoration. Called annatto, this is now a common additive in many processed foods and fruit juices. During the second world war it was distributed in packets which, kneaded into white margarine, made it look like butter. The dye is essentially tasteless. It can be demonstrated easily by rubbing a seed between the fingers.

Blighia sapida

Akee
Ackee

FAMILY: Sapindaceae

HABIT: Tree, to 40 ft.

HABITAT: Cultivated.

NATIVE TO: Tropical Africa.

LEAF: Alternate, 1x even-pinnate; with 6 to 10 pairs of opposite elliptical leaflets, the larger ones to 6 in. long; leaflets barely stalked; not deciduous.

FLOWER: Flowers in stiff, elongate inflorescences; each small, greenish-white, very fragrant.

FRUIT: Hanging, globose (marked slightly into three longitudinal parts), about 3 in. in diameter, smooth surfaced, green turning red, splitting into thirds when ripe.

•Each of the three fruit sections contains a single large shiny black seed surmounting a thick whitish or cream-colored oily stalk (the aril). The cooked fleshy ripe aril is commonly used as a prized food in Caribbean areas. It has a pleasantly nutty flavor. Unripe arils, over-ripe arils, or other plant parts, however, can be severely toxic. The poisoning ("Jamaica sickness"), is technically a severe hypoglycemia, often lethal. The poisonous principle is water-soluble. Cooking in water reduces the toxicity (if the water is discarded), but many people prefer akee fried.

This tree was named in honor of Capt. Bligh of the *Bounty,* though it was introduced into the Caribbean area from Africa by slave traders.

Bougainvillea glabra

Bougainvillea
Paperflower

FAMILY: Nyctaginaceae

HABIT: Shrub or vine to 25 ft.; hedge.

HABITAT: Ornamental; occasionally naturalized.

NATIVE TO: Tropical South America.

LEAF: Simple, alternate, broadly elliptical, narrowing to a point, stalked, not conspicuously hairy; about 4 in. long; variegated with splotches of white in one variety.

FLOWER: Clusters of conspicuous persistent brightly colored bracts (usually 3), with a single small, tubular, short-lived, white or cream-colored flower inside each; bracts crisped on edges, not fading much with time.

STEM: Smooth-surfaced (hairless), bearing sparse short woody spines where the leaves are attached. This species has a looser, more open growth pattern than Bougainvillea spectabilis (see next).

•Bougainvilleas (this species and the next) flower over extended periods but are most spectacular during the dry season. The basic colors are purple or red, but other colors and shades such as orange, brick, pink, crimson, and white have been developed in the numerous horticultural selections of these species. Sometimes more than one color is found on a single plant. Bracts, in evolution not modified much from leaves, are longer-lasting than petals or sepals. Thus bougainvilleas display masses of color more or less continuously over several seasons and are among the most popular of tropical ornamental plants. This species does well and flowers early as a house plant in the north.

Bougainvillea spectabilis

Bougainvillea
Paperflower

FAMILY: Nyctaginaceae

HABIT: Shrub or hedge, to 10 ft.

HABITAT: Cultivated.

NATIVE TO: Tropical South America.

LEAF: Simple, alternate, stalked, broadly elliptical, narrowing to a point; to 4 in. long; more or less hairy especially on the under surface.

FLOWER: As in Bougainvillea glabra above, except purple only, usually fading or changing color somewhat with age.

STEM: Downy-surfaced when young; conspicuous spines, sometimes hooked, where leaves are attached, which allow it to become a climber when there is something to climb on. This species is more compact and bushy, and sometimes coarser, than Bougainvillea glabra. The stems may twine on one another becoming an inch or two thick with age.

•Bougainvilleas are named after Louis de Bougainville (1729–1811), a French navigator, who found them in Brazil and brought the plant back to Europe for cultivation. They readily root from cuttings, grow best in full sun, and withstand drought and heavy pruning well. They are used singly, as hedges, or massed in beds. The bract colors of the two species of Bougainvillea include some of the most intense shades of pigmentation in all of nature. (See also the previous description.)

Brassaia actinophylla (syn. Schefflera actinophylla)

Queensland umbrella tree
Octopus tree
Schefflera
Australian umbrella tree

FAMILY: Araliaceae

HABIT: Tree, to 40 ft.

HABITAT: Ornamental.

NATIVE TO: Australia (Queensland).

LEAF: Palmately compound; 7 to 18 leaflets; each leaflet with a separate short stalk before they join to the leaf-stalk; glossy, leathery, elliptical, somewhat broader beyond the center, usually smooth-edged but sometimes slightly toothed; the whole leaf ("umbrella") 20 to 30 in. across. The leaf stalks may be 2 ft. long before they divide. The stalks of the individual leaflets are joined to each blade with a conspicuous swollen joint. The disk from which they radiate is joined to the leaf-stalk proper with a similar, larger, conspicuous, bent swelling.

FLOWER: Individually small, inconspicuous, reddish-purple; in clusters along conspicuous stiff spreading inflorescence branches; each branch to 3 ft. long.

FRUIT: Small, spherical, dark red-purple, short-stalked, clustered as were the flowers.

STEM: Single (usually) or multiple trunk, sparsely branched.

•The radiating branches of the inflorescence when bearing clusters of fruit, are fancifully like the bumpy arms of an octopus, whence the second common name. This is a common pot plant of hotel lobbies, etc., and is often called Schefflera in horticultural trade. It is widely planted as a landscape tree in the tropics.

Breynia disticha (syn. Phyllanthus nivosus)

Jacob's coat
Snowbush

FAMILY: Euphorbiaceae

HABIT: Shrub or hedge, to 8 ft.

HABITAT: Ornamental.

NATIVE TO: Pacific islands.

LEAF: Simple, alternate, short-stalked, more or less two-ranked along the stem; 1 to 2 in. long; purple-red, white, variegated, or mottled near the top of the plant; green below.

FLOWER: Small, greenish, inconspicuous, on short flower stems among the leaves. Sexes separate but both male and female flowers are on the same plant (monoecious).

FRUIT: Small dark red or black berries, mostly solitary.

STEM: Thin, stiff, often with a zigzag branching pattern; young branches are dark red.

•Several horticultural varieties have been developed with differing leaf colors or patterns of coloration. New stems arise readily from lateral roots creating a spreading pattern of growth. Specimens in the trade are normally grown from root cuttings.

Brugmansia x candida (syn. Datura candida)

Angel's trumpet
White angel's trumpet

FAMILY: Solanaceae

HABIT: Tall shrub or small tree, to 20 ft.

HABITAT: Ornamental.

NATIVE TO: South America.

LEAF: Simple, alternate, stalked, thick, velvety, flannel-like, smooth-edged or slightly toothed or wavy; broadly elliptical, pinnately-veined, to 16 in. long.

FLOWER: Single, large, white, fading to cream; hanging straight down, to 1 ft. long; fragrant (musky); calyx about half as long as the corolla tube, corolla 5-parted, trumpet-like; ending in 5 points that are usually bent back. There is also a double-flowered form.

FRUIT: This plant rarely sets fruit. When it does, they lack the spines that are characteristic of its close relatives.

STEM: Young stems and leaves are hairy, and very brittle.

•Brugmansia candida flowers several times a year. The flowers sometimes have a pinkish or yellowish cast. This and most other species of Brugmansia and the closely related Datura are unmistakable when in flower. Brugmansia candida is a hybrid between two wild species of Ecuador. The "x" between the genus and species names designates its hybrid character. The hanging flowers often look a little wilted or not fully open. The plant is poisonous.

Bursera simaruba

Gumbo limbo
Turpentine
Birch gum
Tourist tree

FAMILY: Burseraceae

HABIT: Tree, to 40 ft. or more.

HABITAT: Deciduous woods, or cultivated ornamental.

NATIVE TO: West Indies, Central America, and south Florida.

LEAF: 1x odd-pinnate; leaflets 3 to 11, oval, pointed, short-stalked, each 2 to 4 in. long; deciduous early in the dry season.

FLOWER: Small 5-parted flowers in inconspicuous clusters.

FRUIT: Globose fleshy pod, sometimes elongate, dark red-purple, slightly ridged, about 1 to 2 in. long, containing 1 to 3 seeds.

TRUNK: Flaky red shiny smooth bark; peeling easily and disclosing green beneath. The trunk is thus usually mottled red and green.

•The bark is the most distinctive feature. Leaves, bark, root, and resin are used in a variety of ways in native medicines. The resin is obtained by injuring the bark. The wood is very light (almost like balsa) and not useful.

A new common name has appeared in recent years in some parts of the Caribbean for this tree. It is now being called "tourist tree" because it is always red and peeling.

Caesalpinia coriaria

Divi divi
Livi divi, etc.

FAMILY: Leguminosae, Caesalpinioideae

HABIT: Tree, to 30 ft.

HABITAT: Native and naturalized.

NATIVE TO: Topical Americas.

LEAF: Alternate, 2x even-pinnate; leaflets numerous, regularly nearly touching to overlapping; each less than 1/2 in. long.

FLOWER: Small, in terminal clusters, white or yellow, pea-like, inconspicuous, very fragrant and attractive to bees.

FRUIT: Small curved, dished, or twisted flat pod with rounded ends, about 1 in. wide; often little longer than wide.

STEM: The trunk and branches are gnarled, with gray bark.

•This is the national tree of Curacao. In sheltered locations, the tree is symmetrical with a spreading mounded top as shown in the illustration. Exposed to the prevailing winds, however, it leans away from the wind, and its top, growing mostly to the lee side, appears to be blown out horizontally in the wind. In this characteristic it can be confused with Crescentia cujete (calabash tree) which may do the same. The pods are a rich source of tannin.

Caesalpinia pulcherrima (syn. Poinciana pulcherrima)

Pride of Barbados
Dwarf poinciana
Flower fence

FAMILY: Leguminosae, Caesalpinioideae

HABIT: Shrub or small tree, to 15 ft.

HABITAT: Ornamental, naturalized; wild.

NATIVE TO: West Indies.

LEAF: Alternate, 2x even-pinnate, "lacy", to 1 ft. long; 6 to 12 pairs of blunt-tipped leaflets each about 3/4 in. long, along the 3 to 9 pairs of secondary axes of the leaves.

FLOWER: Erect clusters of brilliant red, red and yellow, or yellow long-stemmed flowers with long protruding stamens; 5 colored sepals, 5 spreading petals.

FRUIT: Hanging flat legume pods, green, turning brown, 4 to 5 in. long.

STEM: Somewhat thorny along the branches.

•This showy plant brightens rural roadsides and village yards. The globose flower buds are borne on long individual stems. When they open, the fifth petal differs from the other four. It is distinctly smaller, narrow, and curved. This shrub flowers nearly all the time. The sepals drop early. All flowers in a single inflorescence are nearly the same color (all red, or red with a yellow margin, or yellow) but all three kinds may appear on a single plant.

The inflorescences of Caesalpinia pulcherrima are similar in color and structure to those of its close relative, Delonix regia (royal poinciana or flamboyant), but are always distinctly larger in the latter. It can be confused also with Caesalpinia gilliesii (bird-of-paradise shrub) which has light yellow flowers with bright red stamens and lacks thorns.

Calliandra haematocephala (syn. C. inaequilatera)

Red powder puff
Powder puff

FAMILY: Leguminosae, Mimosoideae

HABIT: Shrub or small tree, to 20 ft.

HABITAT: Ornamental.

NATIVE TO: Bolivia.

LEAF: Alternate, 2x even-compound, 5 to 8 pairs of leaflets on each of two secondary axes arising as a pair from the short leaf stalk; the outermost leaflets somewhat the largest (to 3 1/2 in.), often distinctly curved. At the base of the main leaf stalk are two blade-like structures (stipules) where it attaches to the stem.

FLOWER: Small, clustered in large, showy, pincushion-like heads in which the numerous, shiny, long, radiating, brightly colored red stamens tipped with tiny dark globular anthers make the main show.

FRUIT: Flat legume pod with a slightly raised border, to 5 in. long, snapping open at maturity, with the halves then twisting into tight spirals as shown in the illustration.

•This is one of several related and similar species with red, pink, or white "powder puffs" of stamens. Calliandras can be separated from closely related showy Acacias by the raised border on the pod, lacking in Acacia. The largish seeds are well scattered by the sudden almost explosive opening and twisting of the dry pods.

Callistemon speciosus

Bottlebrush
Weeping bottlebrush

FAMILY: Myrtaceae

HABIT: Tree, to 30 ft.

HABITAT: Ornamental.

NATIVE TO: Australia.

LEAF: Simple, alternate, almost stalkless, smooth-edged, narrow-elliptical, about 4 in. long, sharply pointed; bronze green when young, greening with age; speckled with tiny dots.

FLOWER: Dense hanging, showy "brushes" of small, 5-parted flowers, the inflorescences to 6 in. long; scarlet red; consisting of radiating stamens, to 1 in. long, with bright yellow anthers, giving the bottlebrush aspect.

FRUIT: Globose or somewhat flattened, to 1/4 in.; almost stemless, dimpled at the tip, and clustered along the the branch where the inflorescence was earlier.

•The flower clusters develop at the tips of the branches, but the branches continue growing as the flowers open, fade, and set fruit. Thus the mature fruit come to lie clustered on a bare spot on the branch some way back from the leafy tip.

There are some twenty species in this genus of which several are used as ornamentals and are difficult to tell apart from the one described here. A yellow-flowered "bottlebrush" is Melaleuca quinquenerva, the paperbark tree.

Calotropis procera

Crown flower
Giant Indian milkweed

FAMILY: Asclepiadaceae

HABIT: Stiff coarse herb or shrub to 12 ft.

HABITAT: Ornamental, naturalized weed.

NATIVE TO: India, Middle East, East Africa.

LEAF: Simple, opposite, broadly elliptical, pointed, base cleft and clasping the stem, almost no leaf stalk, white-woolly underneath, to 8 in. long; milky sap usually shows if broken.

FLOWER: Clusters of 5-parted flowers, each with thick,purple-tipped white petals, spreading to about 1 in. across; a white or purple "crown" at the center. The flower bud is swollen like a balloon.

FRUIT: A swollen dry pod, 3 to 4 in. long, containing a mass of silk. Fruiting is uncommon in some areas.

STEM: Most of the stems are straight and stick-like.

•The petals look something like a five-armed starfish in outline. Young plants (or the branch tips of older ones) look like a large milkweed with very large purple flowers. Another species, C. gigantea, grows larger overall and the flowers are somewhat larger with mostly bent-back petals and a more prominent "crown" at the center. One variety is entirely white. Crown flower is commonly used in leis in Hawaii. One kind of lei is made with the whole flower; another with just the central crown. The latter look almost as though carved from ivory.

Canavalia maritima (syn. C. obtusifolia)

Bay bean

FAMILY: Leguminosae, Faboideae

HABIT: Vine, mostly on ground.

HABITAT: Seashore, beaches.

NATIVE TO: Polynesia.

LEAF: Alternate, 1x odd-pinnate, with only 3 leaflets per leaf (see illustration); leaflets are wide and slightly pointed, or blunt-tipped, or notched.

FLOWER: Few-flowered clusters sticking up above the vegetation on long, erect stems; composed of purple or pink pea flowers, each about 1 in. across.

FRUIT: Boxy, flattened-but-thick legume pod about 1 in. wide and of various lengths up to 6 in.

•This plant is widespread on tropical shores in both hemispheres, often pioneering by forming long, creeping strands growing across the beach sand. It is an important stabilizer of beach erosion where it occurs, and should not be disturbed. There are many other species of Canavalia in various habitats in the tropics.

Bay bean can be confused with Ipomoea pes-capri and its relatives, or Sesuvium spp. which are also beach runners, but the flowers are very different among these genera.

Carica papaya

Papaya
Pawpaw
Melon tree

FAMILY: Caricaceae

HABIT: Tree, to 25 ft.

HABITAT: Cultivated; sometimes naturalized.

NATIVE TO: Tropical America.

LEAF: Simple, attached in a spiral; long hollow leaf stalks, clasping below; blades large, palmately 7-lobed; lobes pinnately cut; milky sap.

FLOWER: Male: drooping clusters to 3 ft. of many long-stemmed yellow flowers, about 1 in. across. Female: few or solitary, pale yellow, short-stalked from trunk, to 2 in. across.

FRUIT: Pendant, elongate or globose, to 20 in. long, green to orange when ripe, with yellow or orange flesh and numerous black seeds lining a hollow center; to 8 lb.

TRUNK: Soft, greenish, hollow; normally unbranched, with persistent leaf scars in spiral pattern along its length, topped with an umbrella-like terminal cluster of large "tropical" leaves. Fruits are produced close to the trunk among the leaves.

•Papayas are a major crop tracing to pre-Columbian cultivation by Central American Indians. They survive shipping well (unlike many tropical fruits) and are available now commonly in northern markets. Fruits are used in desserts, salads, and juices, or as a delicious breakfast melon. The ripe flesh is sweet and similar to that of some muskmelons. Papaya contains an enzyme used to tenderize meat and in medicine (non-surgical treatment of slipped disc).

In the wild, plants are either male or female. A few male plants may bear occasional perfect (hermaphroditic) flowers which may set fruit. In commercial culture, varieties with perfect flowers have been sought in breeding programs. This makes it unnecessary to cultivate separate male trees for pollination. Plants sometimes begin flowering within three or four months from seed and will fruit for two to three years. The trees are soft-tissued and very rapid growers. Mosaic virus carried by aphids is a very serious disease and has wiped out plantings that were major sources of market fruits in some areas.

The common name, "pawpaw", is also used for a northern tree (Asimina triloba), and this sometimes creates confusion, but the two are entirely different.

Carissa grandiflora

Natal plum
Carissa

FAMILY: Apocynaceae

HABIT: Shrub, to 18 ft.

HABITAT: Ornamental.

NATIVE TO: South Africa.

LEAF: Simple, opposite, leathery, stiff, elliptical, almost clasping; usually with a short-pointed tip; about 2 in. long; densely distributed along the stems; milky sap (see illustration); not deciduous.

FLOWER: Pure white, tubular, waxy; flaring into 5 spreading narrow petals, to 2 in. across; fragrant. Flowers all year.

FRUIT: Plum-like, to 2 in. long, scarlet to purple, containing 6 to 12 papery seeds; glossy or with a waxy bloom on the surface; milky sap, reddish flesh.

STEM: Paired woody forked thorns about 1/2 in. long where the branches divide (see the flower illustration).

•The fruits are edible raw or more commonly used in jellies and preserves. Milky sap occurs throughout the plant, including the edible fruit, thus contradicting the old "rule" that all plants with milky sap are poisonous. Carissa is frequently grown as a hedge. The thorns make it impassable to man and beast alike, and it adapts well to shearing. In full sun, the leaves are often borne in regular ranks down the stems. Numerous horticultural varieties are in the trade. The plant does well where there is salt spray.

Caryota mitis

Fishtail palm
Clustered fishtail palm

FAMILY: Palmae

HABIT: Erect stems, usually clustered, to 40 ft. at maturity.

HABITAT: Ornamental.

NATIVE TO: Tropical Asia.

LEAF: Irregularly 2x pinnate, to 8 ft. long; leaflets large, wedge-shaped, broadening outward, to 6 in. long; appearing jaggedly torn, often obliquely, at the outer edge which is toothed. (The top left illustration is just part of one leaf.)

FLOWER: Male and female flowers are separate but in the same inflorescence. Both kinds are individually inconspicuous. The inflorescence, however, forms a "mop head" of hanging, green, lumpy branches, drying and persistent with age. The larger lumps are the matured fruits (see illustration).

FRUIT: Small, spherical, 1-seeded, slightly fleshy, about 1/2 in. diameter.

STEM: To 4 in. in diameter.

•This "fish-tail" kind of foliage is unique to this genus alone of all palms. The plant fruits from top down at maturity, with "mop heads" appearing successively at lower and lower nodes. Individual stems live about seven years before flowering. The flesh of the fruit is irritant.

Cassia alata

Christmas candle
Ringworm bush
Candlestick bush
Candlebush

FAMILY: Leguminosae, Caesalpinioideae

HABIT: Shrub, to 15 ft.

HABITAT: Ornamental.

NATIVE TO: American tropics.

LEAF: Alternate, 1x even-pinnate, to 3 ft; leaflets in 6 to 14 pairs, larger outward, blunt-tipped (or even sometimes notched), each to 2 in. wide by 5 in. long.

FLOWER: Terminal erect cylindrical "candles", usually exactly vertical, of 6 to 12 brilliant yellow close-spaced or touching flowers which do not open widely.

FRUIT: Pod, about 6 in. long by 1/2 in. wide, green turning brown-black, 4-winged longitudinally with wavy edges.

•Species of Cassia (which total more than 500) contain many active principles and commonly have been employed in various ways for medicinal purposes by native populations in the tropics. The leaves of this species are used to treat skin diseases (whence one of the common names).

Cassia didymobotrya, native to Africa, also called candlebush, is similar but the pods are flat, not winged. Its flower buds are black and the leaflets slightly smaller than in Cassia alata. The second and third illustrations from the top on the right are of this species. Those on the top and left above the white line are of Cassia alata. Those on the bottom and the one in the middle are of Cassia fistula.

Cassia fistula

Shower of gold
Indian laburnum
Golden rain
Golden shower

FAMILY: Leguminosae, Caesalpinioideae

HABIT: Tree, to 30 ft.

HABITAT: Ornamental.

NATIVE TO: India.

LEAF: Alternate, 1x even-pinnate, to 20 in.; leaflets 4 to 8 pairs, elliptical, pointed, short-stalked, often slightly asymmetric, glossy, 3 to 6 in. long.

FLOWER: Large heavy clusters to 18 in. long; each flower with 5 spreading yellow petals, all nearly the same shape, surrounding protruding, upturned stamens and a single style; the flower buds are distinctly globose.

FRUIT: Straight cylindrical pods, 2 ft. or even 3 ft. long by 1/2 in. in diameter; green, turning dark brown or black, with blunt ends and horizontal markings; lasting on the tree.

STEM: Slim trunk with smooth gray bark; upper branches arching.

•The pods have long been known to yield a pulp that can be used as a fairly gentle laxative. Many Cassias shed their leaves just before flowering making the yellow display even more spectacular. The flower mass in this species nearly equals the mass of leaves, and it is a striking specimen when in full flower. Flower buds are usually large and spherical in most Cassias as in this one, a helpful hint in identifying them.

Cassia spp.

Appleblossom tree
Pink-and-white shower tree
Pink shower tree
Shower tree
Senna

FAMILY: Leguminosae, Caesalpinioideae

HABIT: Tree, to 50 ft.

HABITAT: Ornamental.

NATIVE TO: Indonesia.

LEAF: Alternate, 1x even-pinnate, leaflets in 5 to 12 pairs, each obtuse, to 4 in. long. Leaves are deciduous in winter in most locations.

FLOWER: Hugh hanging or lateral masses of pink and white flowers; petals 1 in. long or more.

FRUIT: Narrow cylindrical pods to 2 ft. long, persistent on the tree, divided within by transverse partitions.

TRUNK: Smooth, blue-gray barked, often arching.

•The description above is for Cassia nodosa, appleblossom tree, which is one of several "shower trees". The flower clusters are reminiscent of apple blossoms. This species has also been hybridized with C. javanica to produce a number of horticultural varieties. Cassia grandis, coral shower tree, and C. javanica, another pink shower tree, are similar to C. nodosa but the flowers are variously peach-pink to lavender. Pink shower trees can be confused with Gliricidia (which see) but all Cassia leaves are even-pinnate, while those of Gliricidia are not.

Cassia surattensis (syn. C. glauca)

Kalamona
Scrambled eggs

FAMILY: Leguminosae, Caesalpinioideae

HABIT: Tall shrub or small tree.

HABITAT: Ornamental and naturalized.

NATIVE TO: Southeast Asia.

LEAF: Alternate, 1x even-pinnate with 6 to 10 pairs of rounded leaflets, each about 1 1/2 in. long, nearly clasping the leaf-stalk. The leaflets are sometimes tipped with a short narrow point.

FLOWER: Flattish clusters of orange-yellow, flat open flowers, with nearly equal petals, partly hidden in foliage at branch ends. The three upper petals are somewhat the larger.

FRUIT: Flat brown pods, in clusters, to 6 in. long.

•This is a common ornamental plant in many places. It flowers nearly continuously through the year in some locations, and is planted by public agencies to beautify roadsides. Scrambled eggs takes that name from the appearance of the inflorescences in the contrasting dark green foliage.

Casuarina equisetifolia

Casuarina
Australian pine
She oak
Ironwood
False ironwood

FAMILY: Casuarinaceae

HABIT: Tree, to 100 ft.

HABITAT: Ornamental and naturalized.

NATIVE TO: Australia and Pacific islands.

LEAF: What appear to be long needle-like leaves are terminal branchlets, 1/32 in. in diameter. Actual leaves are reduced to colorless scales in whorls of 5 to 15 on these branchlets.

FLOWER: Sexes separate. Female: small, red; clustered on short branches; ripening into the cone-like fruits. Tiny male flowers elsewhere on the same tree are clustered in terminal greenish catkins.

FRUIT: Small spherical "cones", to 3/4 in long, first green then brown, which eventually shed seed. The "cones" look like diminutive pineapples. Illustrated in the three closeups are male catkins, unripe fruits (still green), and a spray of "needles".

•Rapid growers, casuarinas look like tall drooping pine trees from a distance. They are commonly planted as a windbreak and often found as a strip of forest just behind beaches. The soughing of the branches as wind passes through is a special sound of tropical beaches. Despite appearances, casuarina is neither a pine nor a gymnosperm, but an angiosperm in which the leaves have been reduced to scales (adaptation to dryness). The photosynthetic green branches look like needles, and the fruit something like a woody cone. The wood is dense and deep red in color.

Catharanthus roseus (syn. Vinca rosea)

Periwinkle
Old maid
Madagascar periwinkle

FAMILY: Apocynaceae

HABIT: Perennial herbaceous shrub, to 2 ft.

HABITAT: Cultivated; naturalized.

NATIVE TO: Madagascar, India.

LEAF: Simple, opposite, smooth-edged, short-stalked; blade broadly elliptical, glossy, 1 to 3 in. long, usually with a blunt but distinctly pointed tip. The veins are a lighter color than the blades.

FLOWER: In few-flowered clusters at branch tips; each flower tubular, rose pink, mauve, or white with a pink center, or all white, flaring into 5 maltese-cross-like petals about 1 1/2 in. across, with bristles in the throat.

FRUIT: A pair of cylindrical pods to 1 1/2 in. long.

STEM: May be reddish in the pink-flowered variety.

•This plant has long been used for medicinal purposes by the native peoples of India. More recently, two potent alkaloids, vincristine and vinblastine, have been isolated and purified from it. These are toxic to actively dividing cells in the body and have become one of the mainstays in chemotherapeutic treatment of human leukemias. The plant, although a common ornamental, is highly poisonous.

Cecropia peltata

Trumpet tree

FAMILY: Moraceae

HABIT: Tree, to 90 ft.

HABITAT: Lowlands; occasionally planted as an ornamental.

NATIVE TO: West Indies, tropical Americas.

LEAF: Simple, alternate, long-stalked; blade round in general outline, to 1 ft. across or more but lobed palmately, into 7 to 12 tips, each elongate-elliptical or broadening outward; woolly underneath. Milky sap.

FLOWER: Short spikes of either all-male or all-female flowers; female spike with small yellow flowers.

FRUIT: Succulent, 1 to 2 in. long.

TRUNK: Hollow.

•This fast-growing, soft-wooded tree is occasionally planted for quick shade. The smaller hollow branches can be fashioned into flutes. They are, however, often populated by ants.

The closely related C. palmata, snakewood tree, varies in some leaf details.

Ceiba pentandra

Kapok
Silk cotton

FAMILY: Bombacaceae

HABIT: Tree, to 150 ft. or more.

HABITAT: Rainforest; cultivated as a useful tree pantropical.

NATIVE TO: Probably American tropics.

LEAF: Alternate; palmately compound when mature, with 5 to 7 narrow elliptical leaflets, each about 1 in. wide to 6 in. long, narrowing at both ends; attached in a whorl at the top of a long leaf stalk. Deciduous in the dry season.

FLOWER: Usually crowded in clusters at the branch tips; flowers white or pale pink, 5-parted, about 1 1/2 in. long; petals 5, elongate or curled in, outer side is heavily hairy; styles and stamens protruding.

FRUIT: Large pod, 1 to 2 in. wide by 4 to 8 in. long, containing numerous brown-black seeds covered by whitish soft hairs which originate in the pod wall.

STEM: The trunk flares into several usually narrow buttresses at the bottom and bears numerous short blunt spines (where not rubbed off). The main branches tend to be nearly horizontal.

•The pods look like large milkweed pods and contain a springy cottony mass of floss. This is the kapok of commerce which was once widely used in life preservers before synthetic fibers were available. Each fiber has a waxy coating and is water repellent. This tree is bat-pollinated and produces flowers that are fragrant at night. Kapok trees are excellent for shade with wide-spreading branches. It is commonly found in market places and is displayed on the national crest of Nicaragua.

Cephalocereus royenii

Cactus
Pipe organ cactus
Organ pipe cactus

FAMILY: Cactaceae

HABIT: Erect perennial succulent.

HABITAT: Dry scrub, thorn woodlands.

NATIVE TO: West Indies.

LEAF: None.

FLOWER: Small, nocturnal, scattered in tracts of woolly hair among the spines in a specialized, short section formed at the tip of mature trunks.

FRUIT: Sub-globose, red, about 2 in. in diameter, red-fleshed.

STEM: Branched mostly near the base, vertically fluted, columnar, about 4 to 6 in. in diameter, gray-green, with 7 to 11 somewhat rounded ribs bearing clusters of spines along their full length; spines woody, needle-like, of varying length to 2 in. long.

•The flowering head (cephalum) at the tip of a mature stem, as shown in the left illustration, is almost the same diameter and the same color as the stem below (unlike Melocactus intortus, turk's head cactus). The evolutionary loss of leaves in this cactus, as in most, is an adaptation to a dry habitat. Photosynthesis is accomplished instead by the stems. Fluting is an adaptation providing a greater green surface and hence more photosynthetic area than provided by a simple cylindrical plant stem. The pithy center of the stem is a further adaptation to dryness as moisture-storing tissue.

Cinnamonium zeylanicum

Cinnamon

FAMILY: Lauraceae

HABIT: Tree, to 50 ft.

HABITAT: Cultivated for spice.

NATIVE TO: Ceylon, India.

LEAF: Simple, opposite or clustered, almost stalkless; blade pointed or blunt, glossy, stiff, elliptical, to 7 in. long; 3 conspicuous main veins arise from the base of the leaf; not deciduous.

FLOWER: Inconspicuous greenish or pale yellow, in "stemmy" open clusters about as long as the leaves.

FRUIT: Small (less than 1 in. long), blackish, dry, slightly pointed berries (young berries illustrated).

•Cinnamon, the spice, comes from the dried brown bark of this plant which is now widely dispersed throughout the tropics. Originally, cinnamon production was a monopoly successively of various colonial powers which protected its status vigorously. Typical spice bark is illustrated, at right center, on the branches of a mature tree. In commercial production, young trees are cut back, which causes the roots to send up slender, erect suckers. These are harvested twice a year for their bark.

Citrofortunella x mitis

Calamondin orange

FAMILY: Rutaceae

HABIT: Shrub or small tree.

HABITAT: Ornamental.

NATIVE TO: Southeast Asia.

LEAF: "Simple" (compound with a single leaflet, as indicated by a joint on the leaf stalk), alternate, leathery, glossy, dark green, 2 to 4 in. long. The leaf stalk is distinctly though narrowly winged or margined and the blade is dotted with glands bearing citrus oils.

FLOWER: Solitary, small, white, aromatic, radially symmetrical, 5-parted.

FRUIT: Spherical, 1 to 1½ in. in diameter; divided into 6 sections, very sour.

STEM: Typical citrus thorns are few and short, occurring singly at the side of the bud, or absent.

•Kumquat (Fortunella spp.) has 3 to 5 sections, orange, (Citrus) 8 or more. This, a hybrid between these two genera, is intermediate. The skin of the fruit is loose as in a tangerine. All species of Citrus and closely related genera (see next) are similar in most details. The confusing relationship of the calamondin orange, a common ornamental species, was worked out by colleagues at Cornell, who proved its hybrid nature. The fact that it is a hybrid is expressed by the "x" in the scientific name (see above).

Calamondin oranges can be used like lemons, but because the calamondin is hardier than most citrus species, and is easily grown, it is used primarily as an ornamental instead.

Citrus spp.

Lemon
Grapefruit
Orange
Shaddock
Mandarin orange
Lime
Tangarine

FAMILY: Rutaceae

HABIT: Mostly small trees.

HABITAT: Cultivated.

NATIVE TO: Southeast Asia.

•All species of the genus Citrus are similar in structure, and similar also to Citrofortunella (see above). Variations occur in **LEAF:** size of blade, width of winging of stalk, and whether the edge is smooth or slightly toothed; **FRUIT:** size and color (green, yellow, or orange), number of compartments (segments), thickness of skin, color of flesh (pink or white), and number of seeds per compartment; and **STEM:** abundance and size of spines. Species vary in general size from woody shrubs to small to medium trees.

Different commercially recognized kinds of citrus usually are different species: orange – C. aurantium; lemon – C. limon; grapefruit – C. paradisi; mandarin orange and tangerine – C. reticulata; and lime – C. aurantiifolia.

Citrus seeds moved along the earliest trade routes from Asia into the Mediterranean area. Columbus brought them to the West Indies on his second voyage (1493) from the Canary Islands. Grapefruit originated spontaneously probably on Barbados around 1750 as a mutant or hybrid of the shaddock (C. grandis). Citrus fruits are second only to grapes in tonnage of fruits harvested worldwide.

Some species will grow only in the true tropics. Others benefit from some yearly frost in the life cycle and will not do well where it is absent.

In the illustrations of Citrus spp., the squat tree is a grapefruit, the other a lime. The closeup of leaves is from a grapefruit tree.

Clerodendrum thomsoniae (syn. Clerodendron thomsoniae)

Danish flag
Bleeding glory bower
Bleeding heart
Bag flower

FAMILY: Verbenaceae

HABIT: Twining shrub.

HABITAT: Ornamental.

NATIVE TO: Australia.

LEAF: Simple, opposite, short-stalked, to 5 in. long, blade broadly elliptical, pointed, smooth-edged, deeply veined, slightly rough-surfaced; not deciduous.

FLOWER: Irregularly clustered; calyx strongly 5-angled and inflated, white or purple, about 3/4 in. long, persistent; the opening narrowed; 5 petals and a spray of 4 stamens, protruding from the mouth, deep crimson, short-lived.

FRUIT: A small berry.

STEM: Smooth surfaced, with vine-like young growth at the tip.

•The crimson flower lasts only briefly, but the inflated calyx pouch is persistent. In the commonest kind, the white turns slowly light purple with time. In a cultivated variety, the young calyx pouch is magenta and it does not fade. Danish flag is now occasionally found in the north as a house plant.

There are many other species of Clerodendrum including a number of ornamentals such as pagoda flower (C. paniculatum), and tube flower (C. indicum). The calyx is not as prominent in these.

Clitoria ternatea

Butterfly pea

FAMILY: Leguminosae, Faboideae

HABIT: Climbing vine, to 20 ft.

HABITAT: Naturalized and weedy; occasionally planted as an ornamental; now pantropical.

NATIVE TO: Probably Asia.

LEAF: Alternate, 1x odd-pinnate, about 6 in. long; 5 to 9 leaflets, each 1 to 2 in. long and quite broad in proportion.

FLOWER: Solitary, nearly stalkless, pea-like, deep blue with white or partly yellow markings. There are also entirely white-flowered and double-flowered varieties of this species.

FRUIT: Flat, narrow legume pod, about 1/3 in. wide by 3 to 5 in. long, containing black seeds.

•The plant is rapid-growing and readily adaptable. The flowers, though small and dominated by the foliage, are a striking deep blue, and attract attention to the plant wherever it grows. Butterfly pea serves to brighten weedy roadsides where it commonly spreads.

Other species with lighter blue to white flowers are common tropical weeds as well.

Clusia rosea

Autograph tree
Balsam apple
Scotch attorney
Pitch apple

FAMILY: Guttiferae

HABIT: Tree, to 50 ft.

HABITAT: Seasides, rainforest; occasionally planted as an ornamental.

NATIVE TO: West Indies.

LEAF: Simple, opposite, almost stalkless, untoothed, leathery, thick, to 8 in. long; broadening outward, often with a blunt or almost square end; not deciduous. Only the mid-vein is conspicuous on the leaf blade.

FLOWER: Single, large, heavy, waxy, dramatic; 7-parted corolla 2 to 4 in. across, petals white often with pink spots or streaks; radially symmetrical; surrounding a massive compound stigma at the center.

FRUIT: Green, then brown woody "apple" to 3 in. diameter, cupped in the persistent calyx, splitting into 7 "peels" around a 7-compartmented central column containing dark red seeds.

STEM: The branching pattern (in the open) is strongly horizontal.

•This tree resists salt and is commonly found near seashores though not limited to them. It may start in soil, but commonly starts as does a fig (see Ficus spp.), with germination of a seed in the crotch of a tree. It then forms aerial roots which grow downward over the surface of the host, some embracing it closely (hence the name Scotch attorney). When young, it can be pruned and maintained as a shrub or hedge. The wood is useful for furniture, etc.

Leaves can be "autographed" with a ballpoint pen or anything with a sharp point. This characteristic is widely known and most trees in public places are badly disfigured. Sometimes the petals bear a distinct circular depression or "seal" in their centers which makes the tree even more special in attracting attention. Clusia and Coccoloba. can be separated by leaf veining. Lateral veins are always conspicuous and often colored red in Coccoloba; never conspicuous in Clusia.

Coccoloba uvifera

Sea grape
Kino
Platterleaf

FAMILY: Polygonaceae

HABIT: Shrub or tree, to 30 ft.

HABITAT: Beach, seaside strand; rarely as an ornamental.

NATIVE TO: West Indies.

LEAF: Simple, alternate, smooth-edged, circular or nearly kidney-shaped in outline, to 8 in. across; heavy; borne on a short leaf stalk that extends downward and clasps the stem; veins usually distinctly red; not deciduous.

FLOWER: Closely spaced and radiating at right angles from an inflorescence stem that starts out erect but later hangs as fruits develop; individual flowers small, fragrant,5-pointed; petals flaring, white, with conspicuous white anthers within.

FRUIT: Grape-like in size and final color, with a small terminal point; in conspicuous narrow hanging bunches; single-seeded.

STEM: Mottled gray bark with age.

•The tree becomes straggly and contorted in the seashore wind, but the trunk of an old tree can attain up to 3 ft. in diameter over time. The "grapes" remain hard and green a long time, eventually turning purple one at a time. When fully mature, they become fleshy (soft), have a sweet-sour taste, and can be used for jelly. Coccoloba honey is sometimes available in markets.

Coccoloba can get a start almost in pure sand (illustration lower right) and often serves to stabilize the upper edges of beaches, and is sometimes planted for that purpose. They are also sometimes planted as shade trees.

Cochlospermum vitifolium

Buttercup tree
Virgin Island peony tree

FAMILY: Cochlospermaceae

HABIT: Tree, to 35 ft.

HABITAT: Ornamental.

NATIVE TO: Central and S. America.

LEAF: Simple, stalked, deeply pinnately 5 to 7-lobed (cut half way or more); to 1 ft. across; lobes pointed, small-toothed; conspicuously veined. Deciduous in the dry season.

FLOWER: In terminal clusters; bright clear yellow, about 4 in. or more across, single or doubled (illustrated). Petals in single-flowered varieties are 5 in number, and deeply notched at the apex, surrounding numerous orange stamens and a single style at the center.

FRUIT: Nearly spherical cotton-boll-like pods to 3 in. long, containing a white floss.

STEM: Dark, smooth barked trunk.

•This tree typically flowers at the end of the dry season, usually when the leaves are off, which makes it the more spectacular. Individual flowers look like giant field buttercups if single, or small peony flowers if doubled.

Cocos nucifera

Coconut
Coconut palm

FAMILY: Palmae

HABIT: Tree, to 100 ft.

HABITAT: Cultivated, naturalized, and sometimes as an ornamental.

NATIVE TO: Tropical Melanesia(?).

LEAF: Closely grouped at top of trunk; 1x pinnate; to 20 ft. long; short-sheathed at base when young; leaflets to 3 ft. long and about 200 in number. The lower center-right illustration shows four whole single leaves.

FLOWER: Sexes separate but in the same inflorescence; the club-like inflorescence sheathed as it develops; each inflorescence white, ivory-like, and branched; each branch consisting of a few female flowers at the base and several hundred small male flowers beyond.

FRUIT: The familiar coconut; angular, about 1 to 1½ ft. long by 8 in. through; thick fibrous husk surrounding a large woody-walled nut.

STEM: Trunk is often leaning or bent, scarred (by harvesters), with a swollen base at the soil line.

•Fruiting begins at the fifth year in fast commercial varieties and continues for the life of the tree (70 years or more). Each fruit takes about 1 yr. to mature. A high-yielding tree can produce about 70 coconuts per year. One of the three "eyes" is soft and can be penetrated easily by the developing seedling.

Nuts washed up on tropical beaches produce seaside strands of palms. Besides producing a cool liquid "milk" to drink, the "meat" for nourishment, the foliage for thatching, and a useful oil, the coconut palm would be valued for its ornamental use alone. Arrak is a liquor distilled from the sugary sap; coir is cordage from the fibers; and copra is dried coconut meat. Its name is from the Portuguese for monkey (from the 3-hole "face" on the nuts). This is probably the most basically useful tree in the history of mankind.

The "golden coconut" is a dwarf variety (slow-growing). Its fruits have distinctly yellow husks and it is commonly grown in yards.

Codiaeum variegatum

Croton
Variegated laurel

FAMILY: Euphorbiaceae

HABIT: Herbaceous perennial or shrub, to 15 ft.

HABITAT: Ornamental.

NATIVE TO: Pacific islands and Malay Peninsula.

LEAF: Simple, alternate, thick, glossy, not deciduous; leathery, smooth, stalked, few-lobed or not, not toothed, flat, wavy, or spiral; 3 to 10 in. long, narrow to broadly wide; pointed or blunt, pinnately veined; striped, splashed, or dotted with reds, copper, green, yellow, off-white, etc.

FLOWER: Inconspicuous in terminal spikes. Sexes separate (male illustrated upper right).

FRUIT: Female flowers develop into small, green, spherical fruits marked in three sections.

•The leaf of Codiaeum variegatum, its most attractive part, is extremely variable. Color, leaf shape, and size depend on variety and on individual variation. The leaves of a single plant may vary considerably from top to bottom of a plant, or changing in color as they age, or they may all be about the same. The leaves of Codiaeum variegatum may be confused with those of Acalypha wilkesiana or Graptophyllum pictum.

Hundreds of varieties of Codiaeum have been developed by horticulturalists from the original green-leafed plant (close to that shown in the illustration second from the top on the right), and are now grown commercially all over the world. The range of leaf size, shape, color, and pattern is amazing. Codiaeums are used singly, in borders, or as foundation plantings almost everywhere in the tropics. They may be grown free or pruned or sheared severely. The best color develops in full sun.

Despite its usual common name, this variable species must not be confused with the genus Croton of the same family, some species of which are strongly poisonous.

Coffea arabica

Coffee
Arabian coffee

FAMILY: Rubiaceae

HABIT: Shrub or small tree, to 15 ft.

HABITAT: Cultivated; rarely naturalized or grown as an ornamental.

NATIVE TO: Tropical Africa.

LEAF: Simple, opposite, elliptical, to 6 in. long and glossy dark green; short- or almost un-stalked; blade pointed with smooth or wavy edges.

FLOWER: Small, pure white; appearing singly or in clusters of 4 to 6 along stems where the leaves are attached; each with 5 widely separated spreading petals, each about 3/4 in. long; fragrant.

FRUIT: Single or clustered shiny red berries along stems, about 1/2 in. in diameter, soft and fleshy when ripe, containing two compressed seeds, the so-called coffee "beans".

•Coffee benefits from some shade and is often grown intermixed with suitable shade trees, especially nitrogen-fixers (see Gliricidia). In commercial usage, much of the pulp is removed mechanically, the remainder by "sweating" (controlled decay in water) and the beans are then dried and polished to remove the remaining papery seed coats. A dry process of removing the pulp is used in some places.

Flowering is extended in time and the fruits ripen in scattered fashion. This means selective hand picking must be employed in harvesting. Varieties of C. arabica, grown throughout the tropical Americas and Hawaii, are accepted as the source of the most valuable commercial coffees. Two other species, C. liberica and C. canephora are grown in some parts of Africa. Coffee trees begin bearing at about 3 years, continuing for up to 30 years.

Cordia dentata

White manjack
Jackwood
Clam-and-cherry

FAMILY: Boraginaceae

HABIT: Small tree, to 30 ft.

HABITAT: Savanna, dry roadsides.

NATIVE TO: Tropical America.

LEAF: Simple, alternate, moderately stalked, rough-surfaced, broadly elliptical, somewhat coarse-toothed, to 4 in. long; not deciduous.

FLOWER: Widely spaced in large clusters; each funnel-shaped, usually 5-parted, with small, white, somewhat crinkled and notched petals, about 1/2 in. across.

FRUIT: Conspicuous hanging clusters of fleshy white or greenish white berries, each nearly spherical or slightly elongate, with gluey contents.

STEM: Often multi-trunked.

•The berries look something like those of chinaberry (Melia azedarach), and their gluey contents present opportunities to children. They are not liked by school bus drivers. The wood is sometimes used in furniture making. This tree is occasionally planted for shade or as an ornamental. A tree in full fruit catches the eye.

Cordia sebestena

Geiger tree
Geranium tree

FAMILY: Boraginaceae

HABIT: Small tree, to 30 ft.

HABITAT: Ornamental; dry soils.

NATIVE TO: West Indies and Florida Keys.

LEAF: Simple, alternate, stalked, to 8 in. long; blade somewhat rough to the touch, with smooth or wavy edges, usually narrowing to a point, stiffish, dark green; not deciduous.

FLOWER: In terminal few-flowered clusters; each flower shallow-trumpet-shaped; petals joined, crinkled, crepe-like; orange or scarlet, usually 5-parted; the flower about 1 in. or a little more across. The central flowers open first.

FRUIT: Fleshy green, yellow, or white berries, about 3/4 in.; edible. The pith is gluey.

STEM: Bark of the trunk is light gray.

•The tree flowers, at least a little, all year (illustrated), but is most showy when in full flower at once. It is called geranium tree because the inflorescences are somewhat similar to geranium (Pelargonium) in color and shape. The trunk yields a good furniture wood.

Cordyline terminalis (syn. Dracaena terminalis)

Ti ("tee")
Good luck plant
Dracena

FAMILY: Agavaceae

HABIT: Herbaceous perennial; can attain 10 ft. or more.

HABITAT: Ornamental.

NATIVE TO: East Asia.

LEAF: About 5 in. wide by 2½ ft. long, stalked, the stalk clasping the stem at the base; blade strap-like; venation is pinnate from a distinct midrib although the secondary veins run mostly lengthwise of the blade; blade smooth-edged, green, or red, or green with pink, yellow, or red markings; often with metallic hues.

FLOWER: Spreading clusters to 1 ft. long of hundreds of tiny lily-like flowers; individual flowers are about ¾ in. long, yellow, white, or reddish.

ROOT: A creeping rhizome (underground stem).

FRUIT: Small red berries. Many of the green-leafed varieties rarely fruit.

STEM: Slender, scarred, unbranched or sparsely branched, woody, with leaves in tuft-like clusters at the tips, especially apparent as they get older.

•In the wild species, the leaves are all green. In the commonest cultivars the leaves are red-margined or red splotched with green, etc. The leaves and plants are used in many ways, including medicinal. Commonly, food is wrapped in them to conserve juices while cooking, or they are used like lettuce under salads or other foods. The rhizomes yield a sweet pulp from which native alcoholic beverages were made.

Cordyline is closely related to Dracaena (which see) from which it is distinguished by technical details of the female reproductive structure. The more or less sharp separation between leaf stalk and blade in Cordyline terminalis helps to separate this plant from the many others with elongate strap-like leaves. Ornamentals sold as dracena are often this or another species of Cordyline, or occasionally a species of Pleomele.

Couroupita guianensis

Cannonball tree

FAMILY: Lecythidaceae

HABIT: Tree, to 100 ft.

HABITAT: Planted sporadically as a curiosity.

NATIVE TO: Northern South America.

LEAF: In radiating clusters at branch tips superficially as though palmately compound, but instead simple, alternate, elliptical, smooth-edged or coarsely toothed, short-stalked, pointed, to 1 ft. long; deciduous.

FLOWER: Fragrant, showy; on thick, contorted branching inflorescence stalks that grow directly out of the tree trunk and hang entangled down it. Each flower to 4 in. across, red or red-yellow, with 6 fleshy petals; the center consisting of a thick hook of partly fused stamens below a yellow mat of smaller, densely clustered stamens.

FRUIT: Brown, spherical cannon balls, like husked coconuts, hanging on the inflorescence stems from the trunk, each about 8 in. across, with a thick hard shell enclosing numerous seeds in a soft pulp.

•This tree is bizarre in appearance and unmistakable when in flower or fruit. The leaves are deciduous, new ones coming in bursts of growth often more than once a year. The long, tangled flower stalks grow forth from the trunk over much of its length and occasionally also from the bases of larger branches when the tree flowers. The cannonballs take about 18 months to ripen, when they become unpleasant in odor.

Crescentia cujete

Calabash tree
Gourd tree
Round calabash

FAMILY: Bignoniaceae

HABIT: Tree, to 40 ft.

HABITAT: Savanna; occasionally cultivated as a curiosity.

NATIVE TO: Tropical America.

LEAF: Simple, alternate but mostly clustered along the branches at nodes, almost stalkless; the blade smooth-edged, glossy, elliptical, broadening outward; to 6 in. long.

FLOWER: Directly on trunk and main branches only; very short-stalked; petals yellow, with red or purple veins, lacerate, to 2 in.; distinctly unpleasant odor.

FRUIT: Spherical, smooth, to 1 ft. in diameter; green turning yellow, then brown.

STEM: The trunk is typically short and the branches erect or long-spreading. In windy areas the branches may grow more strongly on the down-wind side of the tree as illustrated here. At a distance Crescentia cujete can be confused with Caesalpinia coriaria (divi divi) which has a similar pattern of growth in windy exposures.

•The flowers are bat-pollinated and each is open only for a single night. The fruits have a thin but hard woody shell which can be used as a bowl when opened. The gourds can also be constricted by tying a band around them when young. Bowls, dippers, and other utensils may be formed in this way or gourds may be used as large rattles if dried intact.

Crossandra infundibuliformis

Crossandra

FAMILY: Acanthaceae

HABIT: Shrub, 3 ft. tall or less.

HABITAT: Ornamental.

NATIVE TO: India.

LEAF: Simple, opposite, stalked; each pair alternating at a right angle with those above and below; blade 2 to 5 in. long, pointed, glossy, elliptical, usually broader before the middle, smooth-edged, wavy surfaced; not deciduous.

FLOWER: Clustered and overlapping in a convex display; orange or red-orange, conspicuously not symmetrical; the petal tube flares open into 5 unequal lobes, an inch or more across, mostly arrayed to one side of the center which is marked by a yellow spot.

FRUIT: Small pod, to ½ in.

•This plant is best adapted to growth in the forest understory, needing shade. The inflorescence matures a few flowers at a time over many weeks. In the tropics, the plant is in flower more or less continuously.

Cryptostegia grandiflora

Purple allamanda
Rubber vine
India rubber vine

FAMILY: Asclepiadaceae

HABIT: Viny shrub, to 20 ft. if not pruned.

HABITAT: Ornamental.

NATIVE TO: Tropical Africa and Madagascar.

LEAF: Simple, opposite, thick, smooth-edged, broadly elliptical, usually slightly broader below the middle, shiny; 3 to 4 in. long, with very short stalks or none.

FLOWER: Lilac or deep purple, trumpet-shaped, open throated, with 5 flaring petals, 2 to 3 in. across. The petals overlap, one edge under, the other over, the next.

FRUIT: Paired, inflated pods, sharply angled and with pointed tips; to 4 in. long.

•Purple allamanda blooms almost continuously. It yields a milky sap which dries to a rubbery substance. It is among the top five plant sources of latex from which natural rubber is made and it was grown for that purpose in the second world war. Cryptostegia grandiflora will grow as a vine if left unattended, but it is usually clipped back to contain it as a shrub. The flower buds are darker in color than the flowers when open, and pointed like a rocket (illustration middle right).

Cryptostegia madagascariensis is similar to this species, but with redder flowers. Cryptostegia grandiflora (purple allamanda) can be confused with Thunbergia grandiflora by its flower, and by common name with Allamanda violacea which is a purple Allamanda.

Cuscuta spp.

Dodder
Strangle weed
Love vine
Devil's gut
Hell weed
Yellow love

FAMILY: Cuscutaceae

HABIT: Tangled parasitic vine climbing on the host plant.

HABITAT: Opportunistic; not choosy about host.

NATIVE TO: Tropical America and southern U.S.

LEAF: Reduced to scales; alternate, inconspicuous, not functional.

FLOWER: Tiny, white, in clusters along the stem.

STEM: Coarsely threadlike, tangling in, and often twining about the host and itself.

•The tangle of dodder vegetation is pale yellow to orange-yellow, usually abundant and quite conspicuous. The plant sets seeds which fall to the ground and germinate in the soil. The seedling arises from the ground as a non-parasitic, free-living plant, but when it finds a host on which to climb, it becomes a true parasite, giving up its leaves and chlorophyll, and henceforth living entirely at the expense of the host to which it is attached by suckers that penetrate the bark. This plant is unmistakable by the structure and especially the color of its stems.

Cycas spp. (and others)

Cycads
Sago palm
(Cycas revoluta)
Fern palms

FAMILY: Cycadaceae

HABIT: "Shrub" to low tree.

HABITAT: Cultivation, mostly; some are native to the American tropics.

NATIVE TO: Tropical and temperate worldwide.

LEAF: Palm-like in most cycads; compound, 1x or 2x pinnate, with long, narrow leaflets; usually sharp-pointed; edges toothed or with thorn-like modifications in some.

FLOWER: Sexes separate. In both sexes, the reproductive parts are usually grouped into cone-like structures that are typically borne erect in the center of the plant among the leaf bases, but the larger may hang.

FRUIT: Both male and female cones may be inconspicuous or diffuse as they mature, but in some, the cones are striking. Illustrated are an erect male cone in Cycas circinalis and a hanging female cone in Dioon spinulosum.

STEM: Plants are slow-growing, often of squat habit, with little apparent stem (the stem is mostly underground); but treelike in a few. The foliage is borne in a basal tuft in the first; at the trunk tip in the latter. The foliage is ornamental in most cycads and they are grown primarily for the foliage effect.

•Cycads are gymnosperms and are among the most primitive of seed plants. The seeds themselves are not enclosed by an ovary wall, and are thus "naked". "Gymnosperm" means "naked seed". Cycads dominated the vegetation at the time of the dinosaurs but angiosperms ("seeds in a vessel") were apparently better adapted than cycads to whatever climatic changes eliminated the dinosaurs. Despite one of the common names and the palm-like appearance of the leaves, cycads are not at all related to palms. Other common genera of cycads include Zamia, Encephalartos, and Dioon. Some can attain immense old age (to 1,000 years).

Delonix regia

Flamboyant
Poinciana
Royal poinciana
Flame tree

FAMILY: Leguminosae, Caesalpinioideae

HABIT: Tree, to 50 ft.

HABITAT: Ornamental.

NATIVE TO: Madagascar.

LEAF: Alternate, 2x even-pinnate, to 2 ft. long, "lacy"; each with up to 20 pairs of tiny leaflets, each about 1/4 in. long. Leaves are usually deciduous in winter and appear at flowering or just after. The leaflets typically close by folding together at sundown.

FLOWER: Conspicuous sprays; flowers open from large buds, the covering scales green on the outside, displaying scarlet on the inside when they open; revealing 5 spreading and flaring petals, the central one white or yellow or both, streaked with red or purple and dropping quickly; the others scarlet or orange, longer-lived; individual flowers to 4 in. across.

FRUIT: Large, nearly straight, green, then almost black pod; woody, sword-like; 1 to 2 ft. long; persistent on the tree after leaf drop.

•This striking legume tree may be confused by common name or its flower with several others. Peltophorum pterocarpum is the yellow flamboyant. In Caesalpinia pulcherrima (dwarf poinciana), the flower is similar though smaller, but the stamens do not stick out as much as in Delonix, and some solely yellow flowers are usually present. The intensity of the red flower in Delonix makes almost all others seem tame by comparison.

The wide branching habit often makes Delonix look umbrella-like at a distance. The seeds from its pods are sometimes used as beads. Its flower is the official flower of Puerto Rico.

Dieffenbachia spp.

Dumbcane
Mother-in-law plant

FAMILY: Araceae

HABIT: Herb, to 8 ft.

HABITAT: Ornamental; common pot plant in the North.

NATIVE TO: Tropical America.

LEAF: Clustered; in bud, tightly rolled like a cigar; stalk long, clasping at stem; blades large, elongate-oval or a bit heart-shaped, tip pointed, midrib conspicuous; variegated with white or yellow in splotches usually related to the lateral veins or along midrib; green predominating in some varieties, not in others.

FLOWER: Rarely seen; convolute spathe, partially fused spadix (see description at Anthurium).

STEM: Thick, green, with leaf scars; typically unbranched. Leaves in a single, terminal cluster (see center top illustration).

•This striking tropical plant takes its common name from the presence of a poisonous sap in all parts which causes an intense burning reaction in the mouth and often makes speech impossible for a while. It can even cause death if the tissues at the back of the mouth swell enough to block breathing. The cultivated Dieffenbachias include many variable species, hybrids, and horticultural selections differing in pattern of variegation and other characters. They are commonly found in hotel lobbies, barber shops, and restaurants.

Dombeya wallichii

Pink ball tree
Hydrangea tree

FAMILY: Sterculiaceae

HABIT: Coarse shrub or small tree to 30 ft.

HABITAT: Ornamental.

NATIVE TO: East Africa, Madagascar.

LEAF: Simple, alternate, stalked, large, 12 in. long; blade almost as broad as long, heart-shaped, coarsely toothed; tip always pointed with occasionally two lateral points as well; surface softly hairy.

FLOWER: Numerous in conspicuous hanging, pink or red balls, to 6 in. in diameter; each flower about 1 in. across, with 5 equal petals crowded against their neighbors, and a 5-branched, protruding yellow stigma.

STEM: Green and somewhat angular when young, becoming smooth and gray-barked with age.

•This attractive African plant, with its large "tropical" leaves and hanging, hydrangea-like balls of flowers, is not only showy, but also not easily confused with anything else. However, the wild species has been improved by plant breeders. Hybridization with another species may have been involved, and cultivated varieties may differ from the description above in details of leaf, size of plant, and size of inflorescence.

Dracaena deremensis

Dracena
Striped dracena

FAMILY: Agavaceae

HABIT: Woody perennial.

HABITAT: Ornamental.

NATIVE TO: Tropical Africa.

LEAF: Stalkless, strap-like, parallel-veined, clasping at the base; long, usually with white stripes along its length, to 2 ft. long by about 2 in. wide.

FLOWER: In terminal clusters; small and inconspicuous; lily-like, dark red outside, white inside.

STEM: Developing slowly and obscured in some by leaves; but in others attaining up to 15 ft. in length as it ages, mostly naked, with leaves clustered at the tip.

•There are several common species of Dracaena and numerous horticultural varieties in the market. In some species, the leaves may be separated into blade and stalk. They may have red, green, or white stripes or splotches, or may be entirely of one of these colors. The sap of some species yields a resin, the so-called dragon's-blood varnish.

Dracaena is separated from the closely related Cordyline (which see) on details of the female reproductive structure. They can usually be separated also by the underground parts which are orange or yellow in Dracaena. Compared with small palms, with which they are sometimes confused, Dracaenas have distinctly slenderer stems.

Epipremnum aureum (syns. Raphidophora aurea, Pothos aureus, Scinadapsus aureus)

Pothos
Pothos vine
Hunter's robe
Variegated philodendron

FAMILY: Araceae

HABIT: Strongly climbing vine, to 40 ft. or more.

HABITAT: Ornamental.

NATIVE TO: Solomon Islands.

LEAF: Simple, alternate, pinnately veined, short-stalked; the mature leaf blades broadly elliptical, heart shaped at base; to 3 ft. long, irregularly torn or incut along the secondary veins; solid green or streaked with white or yellow.

FLOWER: Greenish or yellowish spathe/spadix (see description at Anthurium); the spadix about 6 in. long.

STEM: Green, sometimes streaked, flecked, or splotched with white or yellow; hollow, attached to the host tree trunk by string-like roots (see illustration).

•Compared with the northern houseplant version which rarely gets beyond the juvenile condition with leaves two to three inches in length and uncut, a mature Epipremnum growing "wild" in the tropics is a striking sight (see illustrations). The several species of Epipremnum originated and spread from southeast Asia. The more numerous species of the similar Monstera (which see) originated in the American tropics. They are separated botanically on technical details of female flower structure. The degree of perforation or splitting in both relates to how much light the mature leaf receives. Leaves grown continuously in dense shade are without splits or perforations.

Erythrina crista-gallii

Cockspur coral tree
Coral tree

FAMILY: Leguminosae, Faboideae

HABIT: Medium tree, or sometimes growing as a large shrub.

HABITAT: Ornamental.

NATIVE TO: South America.

LEAF: Alternate, 1 x compound; with 3 elliptical, bright green, stalked leaflets, each to 3 in. long, broader below the middle. There may be also two diminutive blade-like appendages at the base of the leaf-stalk. Leaves not deciduous with flowering.

FLOWER: In typically elongate clusters; flowers bright red, the large petal spread nearly flat, broadly oval, about 2 in. long; the other petals half as long or a bit more, wrapped around a short style and stamen cluster.

FRUIT: Cylindrical pod, somewhat lumpy, to 15 in. long.

STEM: Bearing sparsely scattered spines, like large rose thorns; sometimes absent.

•The species described here and the next are perhaps the two most common of the hundred or so in Erythrina. The flowers of this one are vivid red, of heavy substance, looking something like waxen sweet peas.

Erythrina variegata (syn. E. indica)

Tiger's claw
Coral tree
Indian coral tree
Crab claw

FAMILY: Leguminosae, Faboideae

HABIT: Spreading tree, to 60 ft.

HABITAT: Ornamental.

NATIVE TO: Philippines, Indonesia.

LEAF: Alternate, 1x compound, with 3 broadly oval pointed leaflets, each with two small wings or cusps at the base of the leaf stalk; thin, to 6 in. long; deciduous in the dry season. The terminal leaflet is larger than the two beneath.

FLOWER: Dense clusters, not all opening at once; flower pea-like, to 2½ in. across, bright red; the side petals small, enclosing a spray of 10 out-thrust stamens and a single equally long style, the top petal erect, blunt-tipped, much folded, and strongly curved. Variety alba is similar but white-flowered and less showy.

FRUIT: Cylindrical blackish pod, to 1 ft. long, deeply constricted between the seeds which are large and red-brown.

STEM: Gray-barked, with tiny black unbranched thorns.

•The seeds of some species of Erythrina are colored like those of Abrus (which see), but are larger. The wood of this species is very soft. The clustered flowers look like a paw-full of bloody red claws, hence the first common name listed above.

Euphorbia lactea

Candelabra cactus
Hat rack cactus
Dragon bones
Monkey puzzle

FAMILY: Euphorbiaceae

HABIT: Fleshy, sometimes woody, perennial, 6 to 15 ft. tall.

HABITAT: Ornamental.

NATIVE TO: East Indies.

LEAF: The true leaves are minute, located between the spines in new growth, and soon lost.

FLOWER: Not known to flower (usually the sign of a species very well adapted to its ecological niche).

FRUIT: If there are no flowers, there is no fruit either.

STEM: Fleshy, 3- or 4-angled, dark green, 2 to 3 in. wide, with an irregular white or yellowish marbled stripe down the center of each face; thick brown spines in pairs along the edges; stem becoming gray-barked with age.

•The genus Euphorbia has almost 1,000 species. Many are cactus-like, but most of these, including this one, can be distinguished from true cactus (genus Cactus) by their milky sap. Candelabra cactus is a common ornamental plant throughout the tropics and is sometimes used as a hedge or fence. It propagates readily from the roots and also from broken pieces of the stem.

Euphorbia leucocephala

Pascuita
Christmas bush

FAMILY: Euphorbiaceae

HABIT: Shrub or small tree, to 12 ft.

HABITAT: Ornamental.

NATIVE TO: Central America.

LEAF: Simple, alternate, long-stalked, narrow-elongate, to 3 in.

FLOWER: White, densely clustered around the periphery of the whole plant; each "flower" consisting of five small, white, pointed, petals, regularly surrounding a yellow center with 2 or 3 much larger white "petals" to one side. The latter are modified leaves (bracts), each about 2 in. long, elliptical but broadening outward, sometimes tinged with pink.

STEM: Delicate, gray-barked, much branched.

•A large specimen in full flower looks like a dense snow storm. An example growing at the Botanic Garden, Tortola, is shown on the front cover. Euphorbia leucocephala is closely related to poinsettia (E. pulcherrima). Although differing greatly in size, the colorful bracts are similar between the two, and in both the flowering day is controlled by day length. Both flower around Christmas time in the northern hemisphere.

Euphorbia milii (syn. E. splendens)

Crown-of-thorns

FAMILY: Euphorbiaceae

HABIT: Herb (young) or small shrub, to 4 ft.

HABITAT: Ornamental.

NATIVE TO: Madagascar.

LEAF: Simple, broadening outward; 1 to 1 1/2 in. long, pointed-tipped; clustered on new growth; often a few scattered elsewhere.

FLOWER: Flower heads of red or pink showy bracts; the true flowers within are minute, inconspicuous, and short-lived.

STEM: Woody, green when young, then gray; loaded with perpendicular thorns in pairs; spines are stiff, slender, 1/2 to 1 in. long; outermost branches are 1/2 in. in diameter or less.

•Like most Euphorbias, this species has milky sap. It is a very common ornamental plant, flowering almost continuously. In the tropics it makes a good hedge. The bracts are commonly deep red, but also yellow or pink in some varieties. Bracts are showy leaves modified in size, shape, or color during evolutionary time, and associated with inconspicuous flowers for which they perform the function of attraction. Leaves, even modified as bracts, usually last significantly longer than do petals. Thus, the presence of bracts often means that the plant possessing them remains attractive from the human point of view for a longer time than if it depended on its shorter-lived flowers alone.

Euphorbia pulcherrima (syn. Poinsettia pulcherrima)

Poinsettia
Christmas star

FAMILY: Euphorbiaceae

HABIT: Shrub, to 15 ft.

HABITAT: Ornamental.

NATIVE TO: Central America and Mexico.

LEAF: Simple, alternate, long-stalked, variably coarsely toothed or lobed, with hairy undersurface; 4 to 7 in. long.

FLOWER: The true flowers are inconspicuous within a single whorl or double ruff ("double flowered") of large, showy elongate bracts (see Euphorbia milii), 10 to 22 in. across; red, pink, or white (red and white are illustrated in the middle left photograph).

STEM: Brown, woody, sparsely branching. Stems and leaves have a milky sap.

•Flowering is triggered by day length (short), and under outdoor conditions in the tropics occurs so as to bring poinsettias in the northern hemisphere into full flower around Christmas. Unlike the highly bred houseplant, some tropical outdoor varieties shed their leaves before flowering. The common one that does this is "double flowered" and especially showy, with intensely red bracts. Poinsettias are usually propagated by cuttings. They contain milky sap and some kinds (especially "wild" varieties) may be mildly poisonous to eat.

Euphorbia tirucalli

Pencil tree
Milkbush
Indian tree

FAMILY: Euphorbiaceae

HABIT: Shrub or small tree, to 30 ft.

HABITAT: Ornamental.

NATIVE TO: South Africa.

LEAF: None (reduced to green appendages on new growth; quickly shed – see illustration upper right). The numerous branchlets do the photosynthesis for the plant and impart a shaggy, brushy appearance to it.

FLOWER: Inconspicuous, greenish.

STEM: Trunks with gray, fissured bark; branches fleshy, smooth, bright green, pencil-like or bigger, final twigs loosely whorled, curving outward and upward, blunt-tipped, 4 to 5 in. long by 1/4 to 1/2 in. in diameter.

•Most species of Euphorbia, like this one, contain a milky sap. It is usually mildly irritating to the unprotected skin, but in the pencil tree it is a strong, even dangerous irritant. For the same reason, this species is poisonous, causing severe irritation to the mouth and digestive system if eaten. Not all plants with milky sap are poisonous (see, for example, Carissa), but milky sap is a good warning signal until you know otherwise for sure.

Ficus benghalensis

Banyan tree
Indian banyan
Vada

FAMILY: Moraceae

HABIT: Gigantic tree (in time), spreading; to 100 ft. tall.

HABITAT: Sacred, introduced.

NATIVE TO: India, Pakistan.

LEAF: Simple, alternate, leathery, broadly oval; shiny, stalked, to 8 in. long; not deciduous. Veins conspicuous and mostly connected at the outer ends near the edges of the blade.

FLOWER: Minute and inconspicuous.

FRUIT: Tiny reddish or yellowish fig, single, in pairs, or in clusters of pairs on short stems; globose, about 1/2 in. across, surface slightly fuzzy.

STEM: A strangler fig; – see description at Ficus spp.

•The mature banyan, which usually starts as a strangler fig, eventually produces thousands of hanging aerial roots that descend to the ground from its branches and become new trunks. In this way the banyan extends itself laterally almost indefinitely. One original tree can literally cover several acres in time. Banyans grow rapidly, are often planted in public grounds, and are conspicuous. There is nothing else quite like it. The illustration bottom right is of a young tree, just beginning to spread. Older trees in public parks are usually severely pruned to limit the number of new roots that reach the ground to a few or none. Otherwise the tree may take over the whole park given time.

Ficus elastica

Indian rubber tree
Fig
Rubber plant

FAMILY: Moraceae

HABIT: Tree, to 100 ft.

HABITAT: Shade tree; formerly cultivated for rubber.

NATIVE TO: India and Malaya.

LEAF: Simple, alternate, shiny, leathery, thick, pointed; to 12 in. long; blades uniform in size and shape; lateral veins numerous, exactly parallel, and joined at the outer end; not deciduous.

FLOWER: Minute, inconspicuous.

FRUIT: Tiny somewhat elongate reddish or yellowish; to 1/2 in. across; in pairs, each embedded in the end of a short, thickened stalk.

STEM: This plant has small, thread-like aerial roots and smooth gray bark. Leaves and branches contain a copious milky sap.

•The young leaves are protected in bud by a conspicuous reddish sheath which is shed as the leaf unfolds (illustration second from top on right). Active branch tips look like pointed cigars, and identify this fig easily from other common species. The leaves in one variety are marked with splotches of light green or yellow (upper right).

Ficus elastica is grown as a large pot plant commonly in the north. The milky sap yields raw natural rubber, and was the first commercial source of rubber. Commercial rubber now comes primarily from the rubber tree (genus Hevea).

Ficus lyrata

Fiddleleaf fig
Banjo fig

FAMILY: Moraceae

HABIT: Small tree, to 35 ft.

HABITAT: Ornamental.

NATIVE TO: Tropical Africa.

LEAF: Large, simple, alternate, to 15 in. long; blade wavy, smooth-edged, glossy, short-stalked; unlike most figs the leaf tips are broadly rounded. In size and outline the blade is often shaped like a fiddle with a waist and narrowed neck by which it is attached to the stalk. The veins are prominent and distinctly yellow.

FLOWER: Small, inconspicuous.

FRUIT: Globose fig, solitary or in pairs, about 1 1/2 in. in diameter, mottled white.

STEM: No aerial roots.

•This conspicuous tree with large dark green leaves is becoming an increasingly popular ornamental subject. Among the ornamental figs, its fruits are larger than most.

All species of Ficus are properly called figs. Ficus carica is the true edible fig. It is grown almost exclusively in Mediterranean countries, especially Spain and Italy, and in California. In most species of Ficus, the fruit is of the same general structure as the edible fig, though only 1/2 in. or less in diameter. Each species of fig is pollinated by the activities of a specific wasp. If the wasp is not also introduced with the fig to a new area, the figs there will be sterile, setting no fruit.

Ficus religiosa

Bo tree
Peepul tree
Sacred tree
Banyan

FAMILY: Moraceae

HABIT: Large tree, to 100 ft.

HABITAT: Planted occasionally as an oddity or for religious significance.

NATIVE TO: India.

LEAF: Simple, alternate; leaf stalk long and slender; blade heart-shaped, with tip extended as a narrowing tail-like point often half as long as the rest of the blade; deciduous. The veins are pinnate, conspicuous.

FLOWER: Inconspicuous.

FRUIT: Small dark purple figs, to 1/2 in. in diameter.

STEM: No aerial roots; bark is smooth and gray.

•The leaves rustle in the lightest breeze. The elongate leaf tips and the hanging nature of the leaves aids in the rapid shedding of rain-water. Leaves that dry quickly in the tropics are less susceptible to fungus attack, which may be an evolutionary advantage. This species can be recognized readily by its "drip-tip" leaves. The tips are often more pronounced than illustrated here.

Tradition says that Buddha meditated under a bo tree for six years. This species now is planted at Buddhist temples the world around.

Ficus spp.

Strangler figs

FAMILY: Moraceae

HABIT: Epiphytes (see below), eventually rooting.

HABITAT: Opportunistic, starting in other trees as available.

NATIVE TO: Tropics worldwide.

•The surfaces of trees in the tropics are usually "dirty" with moist dust or organic matter that has accumulated on them. Many of the hundred or more species of Ficus, including some described in this book, start growth from a tiny seed which a bird or other animal (or even the wind) has transported and randomly lodged in a crotch or fissure of bark in an established woody tree.

If the circumstances are right, the Ficus seed will germinate and form a seedling. As the seedling grows, it sends out roots which attach it to the surface of the host tree (but do not penetrate or parasitize the host). The young Ficus plant obtains its mineral nutrients from that "dirty" surface of its host, and its water from rain or dew. Its leaves are fully photosynthetic and, given adequate minerals, moisture, and light, the seedling can grow indefinitely.

Plants living on, but not parasitizing other plants, are called epiphytes. As it grows, a system of roots is formed by the epiphyte which makes a network downward over the surface of the host tree, and eventually reaches the ground. Now the young strangler fig has unlimited access to minerals and a less chancy supply of water. It then usually out-grows its host and envelopes the latter completely as its roots grow and fuse together. Successful roots eventually kill the host and form a self-standing trunk for the fig. These steps are illustrated progressively in the photographs.

Galphimia glauca (syn. Thryallis glauca)

Thryallis
Shower of gold
Rain of gold

FAMILY: Malpighiaceae

HABIT: Shrub, to 10 ft.

HABITAT: Ornamental.

NATIVE TO: Mexico, Guatemala.

LEAF: Simple, opposite, elliptical, stalked, to 3 in. long; blade blunt-ended often with a slight point at the tip.

FLOWER: Clustered densely in spikes at the branch tips, each yellow, 5-parted, to 3/4 in. across; petals separate, spreading flat out, widening abruptly beyond.

FRUIT: A small 3-sectioned swollen globose pod.

STEM: Slender reddish branches.

•Sometimes botanists have special fun. The genus name of this plant was coined as an anagram of another genus (Malphigia) for which its family is named. This species is often planted and pruned to form a sunny hedge along walkways. The conspicuous regularly scattered spikes of bright yellow flowers extending well above the foliage, help in recognition of this plant at a distance. Galphimia vine (Tristellateia australasiae), a related plant, has similar yellow flowers, though paler.

Gardenia taitensis

Gardenia
Tahitian gardenia
Tiare

FAMILY: Rubiaceae

HABIT: Shrub or small tree, to 15 ft.

HABITAT: Ornamental.

NATIVE TO: Polynesia.

LEAF: Simple, opposite or whorled, elliptical, broadening beyond the middle, very short-stalked and somewhat clasping the stem; blades shiny, heavy, stiff; the veins pinnate and conspicuous.

FLOWER: Striking; solitary, tubular, white, fragrant, with 5 to 9 (usually 7) spreading narrow petal lobes, not overlapping when fully open, and a similar number of stamens within; about 2 to 3 in. across; the calyx is 4-lobed.

FRUIT: Large (to 2 in.), globose, ribbed.

•The gardenia flower is similar to that of Jasiminum, but this handsome bush can be distinguished from any species of Jasminum by the fact that the latter has only two stamens. A native plant of the South Pacific, Tahitian gardenias have been used in leis in Hawaii and elsewhere from prehistory. They are becoming increasingly popular as ornamental plants throughout the tropics.

Gliricidia sepium

Quick stick
Madre de cacao
Cocoa shade

FAMILY: Leguminosae, Faboideae

HABIT: Ungainly tree, to 30 ft.

HABITAT: Cultivated.

NATIVE TO: Tropical America.

LEAF: Alternate, 1x odd-pinnate; 4 to 7 pairs of leaflets, each about 1 x 3 in.; pointed more than most legume leaflets; deciduous.

FLOWER: Hanging loose clusters of pale pink or purple-and-white, scentless, pea-like blossoms; looking like lilac blossoms from a distance. The tree flowers mostly after leaf drop.

FRUIT: Pods, 5 to 6 in. long by 1/2 in. wide.

STEM: Often grows in multi-trunked clumps. The young branches are hairy.

•This legume is a nitrogen-fixer and roots very easily. It grows rapidly and is used for living fence-posts, for support for vanilla and peppers, and for shade in cocoa and coffee plantations. It also grows well from seed. Despite the size of the inflorescence, the flowering tree is not a particularly dramatic ornamental and it does not attract much notice because of the subdued nature of the massed color. It often grows in weedy places.

Gossypium hirsutum

Cotton
Upland cotton

FAMILY: Malvaceae

HABIT: Shrubby herbaceous perennial, 3 to 6 ft.

HABITAT: Cultivated and naturalized.

NATIVE TO: This species is from the New World.

LEAF: Alternate, simple, broad, stalked; blade smooth-edged (young) or 3 (to 5) lobed; palmately veined. Leaves are irregularly dotted with small black oil glands.

FLOWER: Not clustered; yellow to white, fading to red-purple; to 3 in. across; petals large, overlapping (like a rose), papery; the fused tube of stamens and style, typical for the family, (see Hibiscus), is short. The petals never open widely.

FRUIT: Ovoid pod, 1 1/2 in. long, beaked, smooth surfaced; borne within persistent bracts; bursting with cotton fuzz. Young and opened pods are shown in the illustration at lower left.

•Cotton, the world's most important fiber plant, comes from several species of Gossypium, and particularly from several commercial varieties of this species. Upland cotton (this one) was grown prehistorically by the Incas and Aztecs. In long-staple upland cotton, the fibers are 1 in. long or more. In short-staple cotton they are less than 1 in. long. The cotton flower in opening looks much like a delicate yellow rose flower.

Graptophyllum pictum

Caricature plant

FAMILY: Acanthaceae

HABIT: Woody shrub, to 10 ft.

HABITAT: Onamental.

NATIVE TO: New Guinea(?).

LEAF: Simple, opposite, stalked; blade 4 to 8 in. long, elliptical, gradually narrowing beyond the middle to a point; smooth-edged, sometimes all green, more typically variegated with white, or less commonly with bronze, red, or yellow; often somewhat glossy; not deciduous.

FLOWER: Few-flowered clusters; each flower elongate-tubular, purple or red, opening into several often flamboyantly curled lips; stamens protruding; each flower about 1 1/2 in. long.

FRUIT: A small pod.

•This plant is easily mistaken at first for the more common Acalypha wilkesiana. The leaves of Graptophyllum pictum are generally smaller, narrower, more pointed, stiffer, and glossier than those of typical Acalypha wilkesiana, and the edges are smooth rather than toothed, but the the leaves of Acalypha wilkesiana vary over wide limits in these characters. Graptophyllum pictum is commonly planted as a hedge or as a foundation screen. It bears clipping well.

Guaiacum officinale

Lignum vitae

FAMILY: Zygophyllaceae

HABIT: Small tree, to 50 ft.

HABITAT: Now mostly cultivated.

NATIVE TO: West Indies.

LEAF: Alternate, 1x even-pinnate; 4 to 6 leaflets; leaflets stalkless, about 2 in. long, broadly elliptical, heavy, often upturned; not deciduous. The major veins spread fan-like from the base of the leaflet.

FLOWER: Clustered, tiny, pale blue, felty, 1/2 in. long by 1 in. across, with 5 radiating narrow petals. Flowers twice a year.

FRUIT: Small, fleshy, yellow-orange, heart-shaped, flattened, about 3/4 in. long. The seed within is elongate and red-and-black.

STEM: A slender trunk, with light brown-gray bark mottled with scales, under a light green dome of foliage.

•The wood of lignum vitae is notoriously dense, hard, and resinous; so much so that it sinks in water. In the days when most tools and much machinery were made of wood, lignum vitae was used for such things as bearings. The resin has been used as an antisyphillis drug. Here is an unusual case of a plant species in which the common name sounds more like Latin than does the scientific name.

Heliconia spp.

Heliconia
Lobster claw
False bird of paradise
Wild plantain
Parrot's plantain

FAMILY: Heliconiaceae

HABIT: Perennial herbs growing in clumps.

HABITAT: Cultivated or in roadside vegetation.

NATIVE TO: South and Central America.

LEAF: Simple, long-stalked, the stalks wrapping around each other to form a tubular false stem as in Musa (banana); blade expanded, often torn along the parallel lateral veins in older leaves, giving a palm-like aspect.

FLOWER: Inflorescence compound, on a stem that grows at the center of the leaves, erect or hanging; each flower cluster with a colored, boat-shaped bract (modified leaf) below. Individual flowers are asymmetric, based on a plan of three.

FRUIT: A 3-parted, globose "berry".

•Many of the species of Heliconia defy easy exact identification. Here the common showy ones are divided into three distinctive groups on the basis of inforescence characteristics.

GROUP I Lobster claw, wild plantain (H. humilis, H. wagneriana (syn. H. elongata), H. caribaea, etc.)

The inflorescence is erect, the bracts broad, in two ranks, touching or nearly so at their base, 3 to 15 in number; red, red edged with green, red or pink grading to a yellowish base, etc., depending on species. The flowers are mostly hidden within the bracts.

The leaf is large, long-stalked, to 8 ft. or more; blades to 10 in. wide.

GROUP II Hanging lobster claw, hanging heliconia (H. pendula (syn. H. collinsiana), H. rostrata, etc.)

The inflorescence is hanging, the bracts broad, arranged in 2 ranks (H. rostrata) or spirally (H. pendula), not touching at their base (stem showing between); 6 to 20, and red (H. pendula) or red with green or both (H. rostrata) depending on species. Flowers are mostly hidden within the bracts.

The leaf is large, long-stalked, to 6 ft.; blades to 1 ft. wide.

GROUP III False bird-of-paradise, parrot's plantain (H. psittacorum, H. latispatha, etc.)

The inflorescence is erect, bracts narrow, arranged spirally, widely spaced, 3 to 20, red or yellow or a combination of these with green (depending on species). Flowers are conspicuous, extending from the bracts.

The leaf is large, to 5 ft. in H. latispatha, smaller in H. psittacorum.

All heliconias bear copious nectar at the base of the bract. Those with erect inflorescences and broad bracts also usually trap water in the boat-like bracts. In many species, the inflorescences are produced down among the leaves and are often overlooked. As cut flowers, in contrast, they are exotic, unmistakable, and highly prized. The bracts are persistent and look almost sculpted in some.

Some heliconias can be confused with Alpinia, but the inflorescence of the latter is brush-like, not two-ranked. The inflorescence of Strelitzia is one-ranked. The Heliconiaceae is very closely related to the banana family (Musaceae).

Hibiscus rosa-sinensis

Hibiscus
Hawaiian hibiscus
Chinese hibiscus

FAMILY: Malvaceae

HABIT: Shrub, to 20 ft.

HABITAT: Ornamental.

NATIVE TO: Asia.

LEAF: Simple, alternate, glossy, broadly oval, narrowing outward to a point; edges more or less coarsely toothed; surface often crinkled or wavy; 3 to 6 in. long; blade sometimes asymmetric at the stalk where the three main veins arise; usually green, but variegated with white and a little pink in one variety.

FLOWER: Solitary, enormous, erect, single or double, 5-parted; red orange, pink, yellow, lavender or white, or combinations; to 7 in. across, with typical center (see below) unless doubled. The flower, whether on or off the plant, will remain fresh for a full day and then wilt.

FRUIT: 5-cavitied pod.

•All species of Hibiscus and some related species have an odd "brush" in the flower center which identifies them. Stamens and style fuse into a column extending from the flower. Numerous anthers branch off part way out, forming the brush. The column terminates in a 5-parted conspicuous stigma. Other tropical genera with this structure include Malaviscus arboreus (nodding or Turks cap). Hibiscus syriacus (rose-of-Sharon) is a common hardy northern species which also shows it.

Plant breeders have "doubled" the Hibiscus flower, selecting for specimens in which some stamens have become modified as petals. In these, the staminal tube at the center is more or less deformed.

Some 5,000 hybrids and cultivars of Hibiscus are found in frost-free gardens worldwide where they flower almost continuously through more than half the year. Hibiscus rosa-sinensis is the official flower of the State of Hawaii. A dye can be obtained from the petals of the red variety, and it will stain clothing.

Hibiscus schizopetalus

Fringed hibiscus
Coral hibiscus
Japanese hibiscus
Japanese lantern

FAMILY: Malvaceae

HABIT: Shrub, to 10 ft.

HABITAT: Ornamental.

NATIVE TO: East Africa.

LEAF: Simple, alternate, coarsely toothed, narrowing outward to a point, to 6 in. long; stalk shorter than blade; surface often somewhat wavy. Some forms have variegated foliage.

FLOWER: Single, long-stalked, hanging; pale red, the central column like that of Hibiscus rosa-sinensis, but the 5 petals bent back, deeply and repeatedly cut and curved into a striking display.

FRUIT: 5-parted dry pod.

•This plant is very much like the commoner H. rosa-sinensis in characteristics except that the foliage is somewhat more open (leaves more distantly spaced), and the branches a bit more delicate. The lacerated petals together with the typical central brush give it away, and separate it from all others with which it might be confused. Like H. rosa-sinensis, the plant flowers freely through more than half the year.

Hibiscus tiliaceus

Tree hibiscus
Mahoe

FAMILY: Malvaceae

HABIT: Sprawling shrub or erect tree, to 40 ft.

HABITAT: Lowlands and tropical shores worldwide; sometimes planted as an ornamental.

NATIVE TO: Tropics worldwide.

LEAF: Simple, alternate, stalked; stalk about as long as blade; blade broadly heart-shaped, 3 to 8 in. across, unlobed, heavy, pointed or rounded, distinctly whitish below; not deciduous.

FLOWER: Solitary or few-clustered; 5 overlapping petals opening lemon or light yellow with a red spot at the base of each, about 5 in. across, fading dull red or purple at day's end; staminal brush (see Hibiscus rosa-sinensis) short, not protruding beyond the petals which remain cup-like, never opening full wide.

FRUIT: An erect, pointed, brown pod opening into 5 cavities as the outer wall dries and splits apart into 5 pointed segments.

•This plant is usually shrubby in its native habitat along shores or stream banks, but can become quite a good-sized tree if planted elsewhere. The flowers are short-lived and scattered in the tree, but a few are almost always open. The pliable bark of this tree was formerly used for a coarse tapa cloth, for cordage, and for caulking boats. The wood is light, like balsa, and can be used for cork. See also Thespesia which has some similar characteristics of leaf, flower, and habitat. The fruits are distinctly different.

Hippomane mancinella

Manchineel
Poison apple

FAMILY: Euphorbiaceae

HABIT: Coarse shrub or usually a tree, to 50 ft.

HABITAT: Savanna, shoreside woods, stream banks.

NATIVE TO: West Indies.

LEAF: Simple, alternate, stalked, glossy, elliptical, with pointed tip, rounded base, and smooth or very slightly incut and toothed (scalloped) edges; 2 to 4 in. long; deciduous. The vein pattern is pinnate with a single conspicuous main vein which "bleeds" a milky sap in young foliage if broken (see illustration).

FLOWER: Sexes in separate flowers but both sexes on the same plant; flowers reduced and individually inconspicuous; males borne along woody twigs a few inches long.

FRUIT: Spherical "apples", about 1 to 1½ in. in diameter, green turning yellow before drop, with the odor of apples; large pithy pulp with a single large, bumpy, woody seed at the center.

STEM: Branching is wide-spreading; bark light gray.

•The milky sap is severely irritant, dangerous to eyes and poisonous if ingested. Columbus recorded its poisonous nature, the first written record of it, after his men had encountered it with unpleasant results. Manchineel has had such a bad name that it has now been nearly eradicated from all of Florida where it formerly grew abundantly. The trees are perhaps not as dangerous as their reputation, but it is wise not to picnic under one, or handle broken vegetation. The sap can cause permanent blindness if gotten into the eyes, and severe burns on the skin elsewhere.

Manchineel can be confused with Cassia xylocarpa, another shore tree with small green "apples". The latter, however, has much smaller, thicker, mostly blunt-tipped leaves and no milky sap. Manchineel can also be confused with Thespesia populnia of the same habitat. Thespesia leaves, however, are distinctly, usually broadly, heart-shaped with a definite U-shaped notch in the blade where the leafstalk is attached. The flowers and fruits are entirely different, too.

Holmskoldia sanguinea

Chinese hat
Cup-and-saucer

FAMILY: Verbenaceae

HABIT: Open large loose shrub, to 30 ft.

HABITAT: Ornamental.

NATIVE TO: Himalayas.

LEAF: Simple, opposite, short-stalked, to 4 in. long; blade elliptical, broader below the middle, pointed; edges toothed; not deciduous.

FLOWER: Clustered; each flower about 1 in. across, composed of a long-lived dish or "Chinese hat" of fused brick-red or red-yellow sepals (obviously 5-parted), supporting a short-lived protruding tube of red petals.

•Holmskoldia is occasionally planted as an ornamental subject. The unusual and long-lasting, though not particularly showy, flowers are the main attractive feature. The shrub, which tends to straggle, can be kept in bounds by close pruning (as illustrated), and is sometimes planted as a hedge. If not pruned, its long branches arch gracefully to the ground.

Hura crepitans

Monkey pistil
Sandbox
Hura

FAMILY: Euphorbiaceae

HABIT: Tree, to 100 ft.

HABITAT: Moist wooded areas; fence rows.

NATIVE TO: West Indies and Central America.

LEAF: Simple, alternate, deciduous; blades elongate, narrowly heart-shaped, hairy, edges smooth or slightly indented, to 10 in. long; veins prominent, pinnate; leaf stalks equal or longer.

FLOWER: Male and female flowers separate, but both kinds on the same plant; female flowers single, small, dark maroon.

FRUIT: Erect "sandbox" (a flattened sphere, 3 to 4 in. across, dry husked, with shallow longitudinal valleys like a pumpkin, outlining 10 to 20 cavities within); green at first, then brown; containing large crescent-shaped seeds.

STEM: Yellow-gray bark with short, squat, fleshy spines (see illustration) where not rubbed off. The trunk is often fluted and buttressed at the base.

•The fruit splits explosively when dry (can be dangerous). The seeds are toxic like those of castor beans (Ricinus) which they resemble. The plant contains an irritant milky sap and the dry, opened fruit was once commonly employed as a container for sand used in drying ink after writing. Hence the first common name above.

Hylocereus undatus

Night-blooming cereus
Queen-of-the-night

FAMILY: Cactaceae

HABIT: Coarse straggling vine on ground or climbing on trees, etc.

HABITAT: Cultivated along walls, etc.; naturalized.

NATIVE TO: Unknown; probably tropical Americas.

LEAF: None.

FLOWER: Single, large, very fragrant, upturned on a thick stem, white, radiate, to 14 in. across; numerous spreading narrow greenish sepals and numerous separate broad white petals (sometimes red or partly red) forming a relaxed cup surrounding a dense cluster of hundreds of yellow stamens and a central, single, conspicuously branched, white style.

FRUIT: Globose, deep pink or red, 4 to 5 in. diameter; edible.

STEM: Green, 3-angled (triangular in section), jointed, to 60 ft. long, climbing by aerial rootlets; wings 1 to 2 in. wide, edges hardened and brownish, undulating, with clusters of short spines about 1 1/2 in. apart in depressions at the joints. Each cluster contains 1 to 5 inconspicuous spines each about 1/8 in. long.

•The night-blooming cereus forms its flowers as lateral growths from the stem many days before they open. These branches elongate to a foot or more and swell at the tip into a large, egg-shaped flower bud. The heavily perfumed flower opens rapidly in the evening and lasts only through one night, drooping and wilting the next morning (as shown in the illustration). The opened flower resembles a very large white water lily.

Hylocereus triangularis is similar, but the stems are smaller, they lack the thin horny edges, the spines are more numerous, and the flowers are smaller. The flowers in both species are bat pollinated.

Hymenocallis littoralis

Spider flower
Spider lily

FAMILY: Amaryllidaceae

HABIT: Coarse bulbous herb.

HABITAT: Ornamental.

NATIVE TO: Tropical America.

LEAF: Clustered at the ground, strap shaped, parallel veined, 3 ft. long or more by about 2 1/2 in. wide, with parallel edges.

FLOWER: Clustered, showy; white, 6-parted, with very narrow spreading petal lobes, each to 6 in. long, and a small white cup ("corona") at the center; stamens delicate, to 2 1/2 in. long, darkening outward.

FRUIT: Tightly clustered swollen pods accompanied by the brown withered petals which persist for a time (see illustration).

•This plant is commonly grown in beds or around buildings. The starkly white flowers are particularly showy against the green background of the leaves, and are strongly fragrant.

Some species of Lilium (lily) and Crinum are similar in vegetation and flower structure but usually have some color other than, or in addition to white in the petals, or lack the distinctive white cup-like corona at the center.

Ipomoea fistulosa (syn. I. crassicaulis)

Bush morning glory

FAMILY: Convolvulaceae

HABIT: Bush, 4 to 10 ft., sometimes recumbent.

HABITAT: Ornamental.

NATIVE TO: Florida and tropical Americas.

LEAF: Simple, alternate, narrowly heart-shaped, sharply pointed, to 6 in. long or more; upper surface smooth, the under surface and leaf-stalk are slightly hairy; veins pinnate and conspicuous.

FLOWER: Usually clustered, tubular, flaring broadly, morning-glory-like, 5-parted, about 3 in. across; pinkish lavender, sometimes with a deeper purple throat.

FRUIT: Globose pod.

•Each flower opens in the morning and lasts only one day. The stems are delicate at first, becoming woody with age. Bush morning glory is easily grown and is a common garden ornamental, although often overlooked in a crowded tropical garden because its delicate lavender flowers seem pale among brighter tropical blooms.

Ipomoea horsfalliae

Prince (Prince's) vine
Kohio vine

FAMILY: Convolvulaceae

HABIT: Woody vine, creeping, or climbing to 40 ft.

HABITAT: Cultivated; naturalized.

NATIVE TO: West Indies.

LEAF: Simple, alternate, thickish, stalked; broadly palmately divided into 3 to 5 lobes; to 4 in. long.

FLOWER: Clustered; 5-parted, bright crimson, rose, or magenta, tubular, to 2½ in. long, the petal tube flaring to about 1 in. across, somewhat plaited; stamens white, sticking out about ½ in. beyond the tube mouth.

FRUIT: Small globose pods (see illustration).

•This species is related to sweet potato, and like it, possesses a large tuber. It has been brought into cultivation and selected for intensely red flowers. Prince's vine was taken to Hawaii from the tropical U.S.by Prince Kuhio, accounting for both of its common names. Another viney species, Ipomoea quamoclit (cypress vine), has similar flowers but the foliage is very different. In it, each leaf is composed of needle-like segments arising pinnately from the midrib.

Ipomoea pes-caprae

Seaside morning glory
Railroad vine
Bay hop

FAMILY: Convolvulaceae

HABIT: Prostrate herbaceous vine.

HABITAT: Beaches and shores.

NATIVE TO: Pantropical.

LEAF: Simple, alternate, broadly elliptical or almost circular; to 4 in. across; sometimes indented at the tip; leathery, thick-fleshy, smooth-surfaced and conspicuously pinnately veined; often with the halves folded upward along the main vein.

FLOWER: Single or few-clustered, tubular, morning glory type, 5-parted, spreading to 3 in. across, rose-purple.

FRUIT: Pod, 1 in. across, splitting into 4 sections at maturity; one seed in each section.

STEM: The purplish viney stems can spread for as much as 60 ft. or more across bare sand, rooting at the stem nodes. What appear to be scattered separate plants are usually connected by underground stems (see illustration top right).

•Ipomoea pes-caprae has tremendously long roots and stems which not only supply water, but anchor it in the sand where it grows. Several other species of this genus also grow on beaches. Ipomoea littoralis has white flowers with a deep red throat and pointed leaves. Ipomoea macrantha has entirely white flowers. The runners of all these beach plants are excellent shore stabilizers. Beach morning glories can be confused with other beach plants such as Canavalia maritima (bay bean), but the flowers and leaves are distinctly different when examined.

Ixora coccinea

Ixora
Maltese cross
Jungle geranium
Flame-of-the-woods

FAMILY: Rubiaceae

HABIT: Shrub, to 4 to 10 ft.

HABITAT: Ornamental.

NATIVE TO: India.

LEAF: Simple, opposite, 2 to 4 in. long; with little or no leaf stalk; blade leathery, sometimes broadening somewhat beyond the middle; shiny dark green; pinnately veined.

FLOWER: Numerous in loosely spherical or mounded clusters; each flower bright red or orange, thin-tubular, to 2 in. long, with 4 petals spreading like a maltese cross. A yellow-flowered variety (illustrated) is occasionally encountered.

FRUIT: Spherical red or black berry.

•Ixora bushes flower nearly throughout the year. The flower buds are almost linear and are protruded somewhat beyond the plane of the mounded inflorescence, a feature helpful in recognition. Ixora flowers are used in one of the more delicate and rarer leis in Hawaii. The tiny flowers are picked and strung individually, and perhaps a thousand of them are needed for one lei. Ixora is often grown as a hedge and stands clipping well. Clipping causes the flower clusters to be smaller and more numerous (illustration bottom right).

Several other species and hybrids of Ixora are in cultivation. A common one, Ixora duffii (syn. I. macrothrysa), differs in its leaves which are much larger, up to 1 ft. long.

Jacaranda mimosifolia

Jacaranda
Green ebony

FAMILY: Bignoniaceae

HABIT: Large tree, to 50 ft. or more.

HABITAT: Ornamental.

NATIVE TO: Brazil.

LEAF: Opposite, 2x finely odd-pinnate; the leaflets tiny (about 1/2 in. long), numerous, uniform, stalkless, pointed, giving a lacy appearance; deciduous.

FLOWER: Clustered; each sky blue, trumpet-shaped, about 2 in. long, tubular, slightly less wide than long when fully opened, opening from dark blue buds.

FRUIT: An angular woody pod about 3 in. long, opening into two lengthwise cavities.

•In many parts of the tropics, most of the flowers of a jacaranda appear after leaf drop, just before the new leaves grow forth. A large fully flowering jacaranda is worth travelling to see.

At first encounter, the pinnate foliage and clustered flowers of a jacaranda may suggest that it is a legume. The open tubular flowers, a few of which are present most of the year, immediately separate jacarandas from most legumes, and the pods have two lengthwise cavities instead of one as in all legumes.

Jasminum multiflorum (syn. J. pubescens)

Star jasmine

FAMILY: Oleaceae

HABIT: Climbing viney bush.

HABITAT: Ornamental.

NATIVE TO: India.

LEAF: Simple, opposite, elliptical, short-stalked, to 2 in. long; more or less downy; each "leaf" is really a single leaflet of a reduced compound leaf.

FLOWER: Clustered, white; with 4 to 9 spreading petals and a persistent calyx of numerous spreading narrow (almost thread-like) teeth.

FRUIT: Small berry.

•There are many species of Jasminum with white or yellow flowers. This species is variable in habit, and grows as a ground cover in some places. It may be unperfumed, or may possess a light, lilac fragrance. The thickish downy leaves and the persistent calyx are good clues to its recognition.

Jasminum nitidum

Star jasmine
Angel wing

FAMILY: Oleaceae

HABIT: Viney shrub.

HABITAT: Ornamental.

NATIVE TO: Admiralty Islands.

LEAF: Simple, opposite, elliptical to heart-shaped, pointed, glossy, dark green, to 3 in. long; pinnately veined. Each apparent "leaf" is a single leaflet of a reduced compound leaf; leaf stalk short or absent.

FLOWER: Clustered; reddish in bud, bud narrowly elongate, opening into narrow, elongate, spreading, fragrant white petals, typically 8 or more, the flower 1 to 1 1/2 in. across; calyx teeth persistent.

FRUIT: Small berry.

•The persistent spreading calyx teeth which remain long after petal fall, are a help to recognition. They are long, very narrow, brownish green, and conspicuous. The flowers are fragrant at night. They may be confused with those of some gardenias (Gardenia spp.) or Tabernaemontana, but are immediately separated by the persistent sepals and somewhat broader petals. Also, all jasmines have only two stamens, and most have pinnately compound leaves while the leaves of Gardenia and Tabernaemontana are simple.

Jasminum officinale

Jessamine
Poet's jessamine
Jasmine
Common jasmine

FAMILY: Oleaceae

HABIT: Climbing vine or viney shrub, to 40 ft.

HABITAT: Ornamental.

NATIVE TO: West China.

LEAF: Opposite, compound, 1x odd-pinnate, with 5 to 9 leaflets; the terminal leaflet is distinctly the largest and most pointed.

FLOWER: Scattered clusters of delicately narrow-tubular pinkish buds and spreading white flowers; the flowers with 4 to 5 spreading petal lobes, and a similar number of green, erect, thread-like calyx teeth below; very fragrant.

FRUIT: Small berry.

•This plant was distributed from its native area in the 16th century and has been widely cultivated ever since. The flowers are a source of essential oils used in perfumes.

The petals of this jasmine are broader than those of the other two species described in this book, and the calyx teeth are shorter in proportion. It is easily confused by flower with Gardenia and others, but the compound leaves and number of stamens (2) help differentiate among them.

Jatropha integerrima

Peregrina

FAMILY: Euphorbiaceae

HABIT: Shrub or small tree to 15 ft.

HABITAT: Ornamental.

NATIVE TO: Cuba.

LEAF: Simple, alternate, somewhat glossy, elliptical or sometimes slightly fiddle- or spear-shaped (lobed), narrowing to a sharp point; not deciduous.

FLOWER: Loosely clustered, radially symmetrical, about 1 in. across; 5-parted; petals broad, intensely deep red, with conspicuous yellow anthers against the red.

STEM: Basically a shrub, this plant can be encouraged to become a small tree with a single trunk, and makes a useful accent subject in gardens in this form.

•The leaves of this species are unusually variable in size, shape, and vein pattern. The blade may be widest below or beyond the middle. The basic vein pattern is pinnate, but the bottommost lateral veins are distinctly stronger than the outer ones. In an extreme form, the leaf can appear palmately veined. Leaves also vary in length, reaching up to 7 inches.

Jatropha multifida

Coral plant
Physic nut

FAMILY: Euphorbiaceae

HABIT: Shrub or small tree, to 20 ft.

HABITAT: Ornamental.

NATIVE TO: Tropical America.

LEAF: Simple, alternate, palmately deeply cut into 8 to 12 smooth-edged or few-toothed lobes, giving the foliage a "lacy" aspect at a distance.

FLOWER: Showy, erect, somewhat flat-topped clusters protruding from the vegetation; individual flowers small, red, 5-parted; borne on persistent swollen red branching stems; anthers a contrasting yellow.

FRUIT: Spherical, slightly 3-angled, about 2 in. long; each section with a single, large, oily, brownish seed; pleasant tasting but poisonous.

•The showy swollen red branching system of the inflorescence persists after petal fall and lengthens the duration of the showy floral display. The common name comes from a fancied resemblance of the inflorescence to the branching pattern of a coral. Like many members of the spurge family (Euphorbiaceae), this species has a milky sap. Its inflorescence is similar to the next (J. podagrica) but the plants are entirely different in habit.

Jatropha podagrica

Australian bottle plant
Gout plant

FAMILY: Euphorbiaceae

HABIT: Single stemmed perennial, to 3 ft.

HABITAT: Ornamental.

NATIVE TO: Central America.

LEAF: Simple, clustered at top of stem; palmately 3 to 5 lobed and veined almost from the center; umbrella-like (stalk attached almost in the middle of the blade); long-stalked; blade to 1 ft. across, crinkled.

FLOWER: Protruding flattish clusters of small showy 5-parted red flowers on swollen red stems, a ring of large erect yellow anthers at the center of each.

FRUIT: Somewhat 3-angled sphere, about 1 in. long, or less, each section at maturity with a large seed.

STEM: The single trunk is knobby, scarred, and distinctly swollen above the soil and below the terminal cluster of leaves. The conspicuous scars represent the places at which leaves were attached.

•Although the inflorescence is much like that of some other species of Jatropha, the swollen single trunk separates this easily from all others. Generally, stems persist longer than petals. The showy swollen stems of the inflorescence in this plant extend the attractiveness of the inflorescences both in space and time.

Justicia brandegeana (syn. Beloperone guttata)

Shrimp plant
Mexican shrimp plant
False hop

FAMILY: Acanthaceae

HABIT: Sprawling shrub, to 8 ft.

HABITAT: Ornamental.

NATIVE TO: Mexico.

LEAF: Simple, opposite, short-stalked; blade pinnately veined, elliptical, pointed, 1 to 3 in. long; smooth edged; rough-surfaced.

FLOWER: Terminal nodding spikes of colored over-lapping bracts (modified leaves), each about half as large as, and shaped much like the vegetative leaves; brick red; persistent. The short-lived tubular, curved flowers are white or pinkish spotted within by small areas of darker color, and protrude from the bracts. There is a yellow-bracted cultivar (illustrated).

STEM: Thin, weak.

•The upside-down, keel-shaped flowers appear from within the bracts one or two at a time, dropping soon. This plant can be used as a coarse ground cover if allowed to sprawl, but it is generally clipped and managed as a small erect shrub or hedge.

Kalanchoe blossfeldiana

Kalanchoe
Life plant

FAMILY: Crassulaceae

HABIT: Herb, to 1 ft.

HABITAT: Ornamental; common pot plant.

NATIVE TO: Madagascar.

LEAF: Simple, opposite, short-stalked, broadly elliptical, to 3 in. long; often reddish-edged and scalloped; somewhat fleshy.

FLOWER: Large erect open sprays of small bright red tubular flowers, each with 4 spreading petals; pointed buds.

•This is common florist material in the north. Some cultivars have flowers ranging in the oranges and yellows. Flowering is induced by short day-length. Some of the Kalanchoes may form small plantlets in the leaf notches; hence the second common name above. Although each flower is tiny, the hundreds massed together and their intense color present a noteworthy display in midwinter.

Kalanchoe flammea from Africa is very similar but slightly larger in character.

Kigelia pinnata

Sausage tree

FAMILY: Bignoniaceae

HABIT: Tree, to 50 ft. or more.

HABITAT: Ornamental.

NATIVE TO: Tropical Africa.

LEAF: Grouped in 3's; each 1x odd-pinnate; with 7 to 9 large leaflets; each leaflet elliptical to somewhat broader outward, to 6 in. long.

FLOWER: Widely spaced on a long, hanging stem; deep wine red, velvety, 5-parted, to 3 in. across, wrinkled, 3 petals up, two down.

FRUIT: A long cylindrical hard woody persistent "club", 1½ to 2 ft. long by up to ½ ft. in diameter, not opening; hanging on long "strings" in and below the foliage.

•This species is sometimes planted as an oddity. The flowers curve and extend upward from the hanging stem bearing them, giving the impression of defying gravity. Though large, they are not particularly showy. They open at night, dropping the next morning, and they have a fetid odor. The fruits have no peaceful use and, sometimes growing at head height, are a danger to passers by. Still, Kigelia is unusual.

Lantana camara

Lantana
Shrub verbena
Yellow sage

FAMILY: Verbenaceae

HABIT: Shrub, to 8 ft.

HABITAT: Ornamental; naturalized.

NATIVE TO: Tropical Americas.

LEAF: Simple, opposite, broadly elliptical, elongate to heart-shaped, rough or crinkle-surfaced, hairy underneath; edges small-toothed or scalloped; to 5 in. long; not deciduous.

FLOWER: Mounded clusters about 2 in. across, of small, 4-parted, somewhat bilaterally symmetrical tubular aromatic flowers, flaring into 4 or 5 spreading lobes; opening yellow, aging orange or red. The inflorescence opens from the center outwards.

FRUIT: Small black or blue-black berry.

STEM: Distinctly square and often prickly.

•In one cultivar, the flowers are pure white. In others, yellow or lilac predominate, either changing in color with time, or not.

Lantana montevidensis (syn. L. sellowiana), a trailing viney plant often used as a ground cover, is similar but with lavender flower clusters, not changing color with time. Some species of Lantana (including L. camara) go wild and weedy under some circumstances and become a problem because they are poisonous to grazing livestock.

Latania loddigesii

Latan palm
Blue latan
Fan palm

FAMILY: Palmae

HABIT: Tree, to 40 ft.

HABITAT: Ornamental.

NATIVE TO: Mauritius Island.

LEAF: A large, fan-like palm frond of distinctly bluish cast; long-stalked; blade to 5 ft. long and nearly as broad, divided into numerous radiating, narrow, sharply pointed segments; each segment separated from the next to about half way back from the outer edge.

FLOWER: Complex, branching, single-sex inflorescences; both kinds on the same plant. Flowers small, inconspicuous; the female larger, ripening into the fruit.

FRUIT: A globose, fibrous structure with 1 to 3 seeds within.

STEM: Slow-growing, grayish trunk ringed with leaf scars.

•The family of palms (Palmae) is large, diverse, and economically important in many ways (food, fiber, shade, building, etc.). The many hundred genera of palms can be roughly divided into two main kinds: the more numerous "feather palms" (leaf blades pinnately compound), or the fewer "fan palms" (leaf segments palmately arranged). The latan palm is the only example of the many fan palm species to be included in this book.

Leucaena glauca (syn. L. leucocephala)

Tan tan
West Indies mimosa
Wild tamarind

FAMILY: Leguminosae, Mimosoideae

HABIT: Shrub or small tree, to 30 ft.

HABITAT: Roadsides, open fields, woodlands; sometimes cropped.

NATIVE TO: American tropics.

LEAF: Alternate, 2x even-pinnate; to 1 ft. long; leaflets 10 to 20 pairs, each about 1/2 in. long.

FLOWER: Small globose pincushions of radiating stamens, about 1 in. in diameter; yellowish or white.

FRUIT: Clusters of flat green, then brown, pods, each to 6 in. long or more, with numerous shiny dark brown seeds in a ladder-like linear series within.

STEM: Has no thorns, unlike many similar woody legumes.

•The pods are persistent on the tree and reluctant to open. The seeds are chocolate brown, glossy, with darker markings. They are commonly strung into necklaces.

This species is a nitrogen-fixer, and therefore a good soil builder. It grows readily on poor soils and is potentially a good forage plant, too. Livestock like it. Unfortunately, it also contains an alkaloid (mimosine) which causes loss of long hair (mane, tail, etc.) in horses, cattle, and swine and serious loss of condition will follow if the plant is consumed in moderate amount continuously over a period of time. Leucaena may also be grown to provide quick shade for another crop such as coffee.

Malaviscus arboreus

Sleepy mallow
Nodding hibiscus
Turk's cap
Wax mallow

FAMILY: Malvaceae

HABIT: Variable shrub or a small tree.

HABITAT: Native and cultivated.

NATIVE TO: Tropical Americas.

LEAF: Simple, alternate, typically short-stalked; blade narrow, and coarsely toothed, with the 3 or 5 principal veins arising from the point where the stalk is attached; usually densely hairy underneath.

FLOWER: Single, hanging, about 2 in. long; composed of a narrow or slightly spreading tube of 5 crepe-like, overlapping petals surrounding a slightly protruding staminal tube (see Hibiscus rosa-sinensis).

FRUIT: A berry, later drying and separating into 5 parts.

•The flowers of Malaviscus arboreus give the appearance of unopened Hibiscus flowers. This species is unusually variable in a number of its characteristics. Commonly a shrub, it can sometimes be vine-like, or even grow into a small tree. The leaves may be smooth-edged or toothed. The flowers vary in the degree to which the petals open. Cultivated varieties may have yellow or pink flowers. Its sap is copious and like mucilage.

Mangifera indica

Mango

FAMILY: Anacardiaceae

HABIT: Majestic rounded or spreading tree, to 90 ft.

HABITAT: Cultivated.

NATIVE TO: India.

LEAF: Simple, alternate, clustered at branch tips; narrow, elongate-pointed, glossy, short-stalked, stiffish; to 12 in. by 3 in.; edge wavy; blade pale reddish when young; not deciduous.

FLOWER: Conspicuous, mostly erect, open pyramidal clusters (8 in. long) at branch tips, composed of thousands of tiny, yellowish, hairy flowers on purplish stems.

FRUIT: Hanging, oval, often lop-sided, to 6 in. long; green, often with darker streaks, ripening purple-brown or yellowish (depending on variety).

STEM: The wood and bark are resinous.

•Individual trees sometimes flower in sectors, part of the tree at a time. Ripe fruits are thin-skinned and contain a single large stone. The orange flesh is sweet, eaten raw or used for juice. Even green mangos yield edible pulp. They are used in chutney, jams, jellies, etc. Plants usually bear 3 to 5 years after planting. Numerous selections have been made to reduce the stringy fibers in the fruit and the turpentine odor, but fruits in local markets may still be close to the wild type. Mangoes are of great importance as a fruit in the tropics, second only to bananas.

Mango is in the same family as poison ivy. A person highly sensitive to poison ivy should be careful of mangos as any part of the plant or fruit may cause a reaction.

Manihot esculenta (syn. M. utilissima)

Cassava
Yuca
Manioc
Tapioca

FAMILY: Euphorbiaceae

HABIT: Herbaceous shrub, to 12 ft.

HABITAT: Cultivated for food.

NATIVE TO: Tropical Americas.

LEAF: Simple, alternate, large, long-stalked, palmately divided almost to the stalk into 5 to 9 radiating leaflets; shape varying with culture and variety.

FLOWER: Small, greenish, numerous, in terminal clusters; males and females separate but both sexes on the same plant.

FRUIT: Small (about 1/2 in.), 6-angled, and winged slightly.

STEM: Soft when young, with milky sap; becoming woody with age. Older stems show persistent leaf scars.

ROOT: Clustered, fleshy; each conical or irregular, up to 2 ft. long and as much as 4 in. in diameter.

•The roots are white-fleshed, and rich in carbohydrates, but also rich in a cyanide-producing compound. The poisonous principle is destroyed by heat such as in boiling, roasting, or steaming. The starchy contents of the roots can be extracted by grating, heating, drying, and grinding into a flour. Mashed tubers can be baked in flat "cakes". Small dried balls of starch that accumulate when roots are mashed, kneaded, and washed constitute the tapioca of commerce.

Commercial cassava plants are the result of hybridization and selection over several thousand years. Plants will grow for several years if not harvested and were used by early tribes as food reserves over difficult times since the roots, once grown, remain available in the ground more or less indefinitely without spoiling.

Manilkara zapota (syn. Achras, Sapota)

Sapota
Sapodilla
esapple
Mesple
Naseberry

FAMILY: Sapotaceae

HABIT: Tree, to 100 ft. when old.

HABITAT: Cultivated in Central America, Florida, and the Caribbean.

NATIVE TO: West Indies and Central America.

LEAF: Simple, alternate, clustered at twig tips; leathery, elliptical, dark green, shiny upper surface, about 6 in. long by 3 in. wide; not deciduous.

FLOWER: Small, white, bell-shaped, inconspicuous (see illustration).

FRUIT: Globose, 2 to 3 in. across, green then red-brown; thin skin, somewhat rough surface, short stem; some erect, some not, containing a sweet yellow-brown granular, translucent flesh and large, black, glossy seeds.

•The plant, including the leaves and the the young fruits, is well provided with a milky sap (see scratch on fruit surface in illustration). This is non-toxic. The fruit is used raw or in preserves. The milky latex from the trunk, after boiling, yields the chicle of commerce, used in chewing gum. The trees are usually tapped every three years. The timber is hard, dense, and durable.

Mammee apple (Mammea americana), growing wild in the American tropics, bears fruits of similar size, shape, and color. They also are edible. The two trees can be separated by the leaves which are distinctly shorter and wider in Mammea.

Medinella magnifica

Medinella

FAMILY: Melastomataceae

HABIT: Shrub, to 6 ft.

HABITAT: Ornamental; pot plant.

NATIVE TO: Philippines.

LEAF: Simple, opposite, stalkless, oval with a pointed tip; to 1 ft. long; veins conspicuous; not deciduous.

FLOWER: Inflorescence terminal, hanging, to 1 1/2 ft. long; consisting of large paired or whorled pink bracts (modified leaves) above a much-branched cluster of small purple-pink or coral-red flowers with purple stamens and yellow anthers. The flower bud looks something like an acorn in shape before it opens.

STEM: 4-angled or somewhat winged.

•This plant grows as an epiphyte (see Ficus spp.) in its native rain forest habitat. In leaf and flower, it is considered among the choicest of greenhouse plants. The leaves are veined in a pinnate pattern, but the veins arising near the leaf stalk are conspicuously the strongest. This pattern of veining, shown in the illustrations, is unusual and characteristic of the family to which Medinella belongs.

Melaleuca quinquenervia (syn. M. leucadendron)

Paperbark tree
Punk tree
Cajeput
Australian tea tree

FAMILY: Myrtaceae

HABIT: Landscape ornamental.

HABITAT: Tree, to 80 ft.

NATIVE TO: Australia and western Pacific islands.

LEAF: Simple, alternate, or almost opposite, short-stalked, to 8 in. long; blade thickish, stiff, narrowly elliptical, dull green, pointed at both ends, usually with 5 main veins running lengthwise.

FLOWER: Inflorescence a whitish or pale yellow "bottle-brush" composed of radiating stamens in bundles of 5, organized like brush bristles. The rest of the floral parts are small and hidden.

FRUIT: Clusters, about 6 in. long, on outer branches; each fruit stemless; a small, dry, brown-black rosette.

STEM: Bark light brown, sometimes almost whitish, peeling in layers; spongy beneath.

•The leaves are distinctly aromatic. An unusual feature of flowering is the fact that the terminal bud of the flowering stem starts growing again after the flowers die and the fruits begin to form. The fruits thus come to appear as clusters along leafy branches back a distance from the tip. The bottlebrush tree (Callistemon) is similar in some respects but the bark separates these two easily. The copious papery bark attracts attention to the tree.

Melocactus intortus

Turks cap cactus

FAMILY: Cactaceae

HABIT: Perennial.

HABITAT: Dry open scrub lands; sometimes planted as an ornamental.

NATIVE TO: West Indies.

LEAF: None.

FLOWER: Borne on a terminal, unribbed, reddish knob or trunk of smaller diameter (about 4 in.) than below. Short-lived small pinkish flowers appear at the upper end of this knob intermittently.

FRUIT: Elongate, globose, stemless, red-purple, to 1 in. long.

STEM: Single trunk; thick, squat, barrel-shaped, to 1 1/2 ft. across by 3 ft. tall; with 14 to 22 longitudinal ridges, each about 1 1/2 in. high, bearing needle-like spines 1 to 1 1/2 in. long in groups along the length.

•This is a distinctive plant in the natural tropical flora where it occurs and is sometimes planted in cactus collections or dry rock gardens (the last illustration is of a plant in the wild). Despite its characteristic and distinctive appearance, the short squat plants are usually mostly hidden from casual view among other taller vegetation in the dry locations where it grows.

Merremia tuberosa (syn. Ipomoea tuberosa)

Wood rose
Yellow morning glory

FAMILY: Convolvulaceae

HABIT: Woody climbing weedy vine.

HABITAT: Weed; occasionally cultivated for the "rose".

NATIVE TO: Tropical America.

LEAF: Simple, alternate, smooth; blade palmately deeply cleft into 5 to 7 lobes; 3 to 8 in. wide by about 6 in. long.

FLOWER: Flaring, yellow, morning-glory-like; followed by the wood "rose"; the flower buds are small, green, swollen, and pointed.

FRUIT: The calyx closes again over the floral parts after the flower has been fertilized. Over several weeks to months it swells into a sharply pointed, large, lighter green or yellowish "bud" much larger than the original flower bud, then reopens broadly and persists around the seed pod, darkening and spreading to 1 1/2 in. across and forming the "wood rose", with the spherical papery inflated seed pod at the center.

STEM: Smooth, hairless, sometimes with viney twistings.

•The wood rose looks like a small rose with petals delicately carved from a medium brown wood, surrounding a hollow brown spherical body of similar texture in the center. This attractive, delicate, and long-lasting fruiting structure is prized for dried floral displays or accent. This species is closely related to the morning glories (Ipomoea), differing in flower color and some other important details of the female reproductive structure. Like many morning glories, it has a large edible tuber.

Messerschmidia argentia (syn. Tournefortia argentia)

Beach heliotrope

FAMILY: Boraginaceae

HABIT: Shrub or small tree, to 40 ft.

HABITAT: Shore locations, open vegetation.

NATIVE TO: Madaegascar, Africa, Australia, and Pacific Islands.

LEAF: Foliage somewhat silvery; leaf simple, alternate, large, short-stalked; blade broader outward, covered with a fine gray-green hariness.

FLOWER: In open clusters; flowers numerous along one side of copiously branching, curving inflorescence branches, tiny and inconspicuous, tubular, with 5 small whitish petal lobes.

FRUIT: Small greenish berries, each with a black dot at its tip.

•This is one of a small number of plants able to thrive at the upper edges of exposed ocean beaches and it may be planted as an ornamental at tropical beach-side resorts. Its somewhat lighter-colored foliage usually makes these plants stand out from other vegetation at a distance.

Monstera deliciosa

Monstera
Split-leaf philodendron
Ceriman

FAMILY: Araceae

HABIT: Climbing vine, to 30 ft.

HABITAT: Ornamental.

NATIVE TO: Tropical America.

LEAF: Simple, alternate, to 3 ft. wide; heart-shaped at the blade base, solid and smooth edged when young, becoming regularly pinnately lobed when older, with elongate holes developing between the lobes, often in 1 or sometimes in 2 series.

FLOWER: Spathe (see Anthurium) creamy, to 1 ft. long; spadix to 10 in. (immature spathe illustrated).

FRUIT: The matured spadix constitutes an edible compound fruit (whence the species name).

STEM: The viney stem is attached to the host tree by string-like roots as it climbs.

•The veins of the leaf blade do not run lengthwise of the blade, but branch from a midvein, and the leaf stalk is winged. These are useful features for distinguishing Monstera from some others. Monstera is often called philodrendron, but is distinguished from the true Philodendron genus in the following way: The point of attachment between the blade and leaf stalk is sharply bent (geniculate) in Monstera, but not sharply bent in Philodendron. Monstera is also easily confused with Epipremnum, and like the latter, has juvenile leaves that are smaller than the mature leaves and without perforations or splits. All these genera have been subject to much confusion of names, both common and scientific, in the past.

Morinda citrifolia

Pain killer
Indian mulberry

FAMILY: Rubiaceae

HABIT: Shrub or small tree, to 20 ft.

HABITAT: Naturalized, shore-side woods, or moist forest.

NATIVE TO: India.

LEAF: Simple, opposite, glossy, elliptical, to 10 in. long by about 4 in. wide; leaf stalks short; stipules (flaps of leaf tissue at the joint between leafstalk and stem) present; lateral veins prominent, curving.

FLOWER: In "heads", flowering outward. Flowers small, white, tubular; petal lobes about 1/2 in. long.

FRUIT: Ovoid, lumpy, compound, about 2 in. in diameter; individual segments outlined, each also with a raised central scar, becoming light green to yellow and soft when ripe.

STEM: The wood is soft and brittle.

•Despite appearances, the fruit is not edible. The flowers yield a red dye and the root a yellow dye. The ripe fruit can be confused with Annona or Artocarpus species (which see). Morinda fruit is softer and smaller than either of these. The name "pain killer" comes from the practice of wrapping leaves around arthritic joints to relieve pain.

Moringa pterygosperma (syn. M. oleifera)

Horseradish tree

FAMILY: Moringaceae

HABIT: Small tree, to 30 ft.

HABITAT: Ornamental; and grown for its edible parts.

NATIVE TO: North India.

LEAF: Opposite, 2 to 3x odd-pinnate; leaflets small, broadly oval and often broader beyond the middle, blunt-tipped, about 1 in. long.

FLOWER: Clustered, fragrant, each about 1 in. long; with ten creamy white petals, 9 bent down, 1 erect.

FRUIT: Triangularly fluted pod 3/4 in. thick, about 1 to 1 1/2 ft. long, containing regular rows of dark brown winged seeds.

STEM: The bark is light colored and corky.

•This tree might be classed as a legume on first glance. The triangular pod quickly separates it, as do details of flower structure. The young pods and seeds are edible. The tree fruits all year in most areas. Roots are pungent like horseradish, and "oil of Ben", used by watchmakers and in cooking, is obtained from the seeds.

Musa x paradisiaca

Banana
Plantain

FAMILY: Musaceae

HABIT: Treelike herb, to 30 ft.

HABITAT: Cultivated.

NATIVE TO: Asia.

LEAF: Blade about 1 by 4 ft., elongate, soon wind-torn along pinnate veins, thus becoming palm-like; leaf stalks long, clasping, forming a hollow "trunk" (see picture detail of cut stem).

FLOWER: Single elongate terminal but hanging inflorescence; sexes separate, females (illustrated) formed first; individual flowers small, 5-parted, in 2 rows of 8 to 15, surrounded by a large dull maroon-colored bract (modified leaf).

FRUIT: Female flowers ripen into a "hand" of bananas. Further outward (down the hanging inflorescence), the flower ovaries are not functional, the flowers are solely male, and abort early.

•The common name, banana, is an African word. The "x" in the scientific name indicates a hybrid of two species. Several other species of Musa and many hybrids also yield useful bananas or plantains, and these reach tropical markets under numerous local names.

Most male flowers do not produce functional pollen, and fruits develop without pollination in most bananas. Thus no seeds develop. Each banana plant flowers only once. The sole growing point is used up thereby, and the plant dies. Bananas flower at 1 1/2 yrs. when the growing point (until then inside the "trunk" at ground level) stops making leaves, starts growing up through the hollow "trunk", comes out the top, arches over, and commences flowering. The hands grow fingers-up in banana, fingers-down in most plantains. The colored bracts which conspicuously enclose the swollen terminal bud when it is producing flowers, are soon shed, but new ones form and the terminal bud keeps growing a long time. Bananas may sucker and are propagated vegetatively. Plantains are starchy varieties of bananas. Eating one uncooked is like eating a raw potato. Bananas have many cultivars, only a few of which ever reach northern markets.

Mussaenda spp.

Mussaenda
Red flag
Buddha's lamp
Ashanti blood

FAMILY: Rubiaceae

HABIT: Shrub, sometimes twining and climbing to 20 ft.

HABITAT: Ornamental.

NATIVE TO: Tropical Africa and western Asia.

LEAF: Simple, opposite; blade oval, narrowing outward, almost stalkless, conspicuously pinnately veined, pointed, hairy beneath.

FLOWER: Each conspicuous "inflorescence" is really only a single complex flower. The loosely clustered, red, pink, or white structures that look like small poinsettia bracts (modified leaves) are actually greatly enlarged sepals (note that they are parallel-veined as compared with the pinnate venation of the true leaves). Often all, but sometimes only one or a few of the sepals are enlarged like bracts. The small central flower-like structure consisting of petals, is tubular, white or orange-yellow, with 5 flaring petal lobes, about 1/2 in. across (illustrated).

FRUIT: A small berry.

•Several closely related species have received horticultural attention in the Philippines and elsewhere, resulting in a number of handsome cultivars with showy inflorescences ranging from pure white through pinks and purples to a true red. The petal tube drops early, leaving a small circular scar at the center of the bract-like sepals. The latter persist, and extend the time during which the shrub is showy. The nearly palmate veining of the sepals contrasts with the clearly pinnate (though curved) veining of the foliage. The larger sepals are nearly half as large as each leaf blade. In the photographs, the purple-flowered ones are of M. alicia; the white ones are M. phillipica.

Myristica fragrans

Nutmeg
Mace

FAMILY: Myristicaceae

HABIT: Tree, to 80 ft.

HABITAT: Cultivated (Grenada especially); rarely naturalized.

NATIVE TO: East Indies.

LEAF: Simple, alternate, short-stalked; blade dark green, leathery, glossy, smooth-edged, elliptical, to 5 in. long; not deciduous.

FLOWER: Male and female trees are separate; female flowers (illustrated) are solitary, small, tubular, somewhat fleshy, white or pale yellow, aromatic, barely opening; no petals.

FRUIT: Spherical, yellow, fleshy; to 2 in., drying and splitting open longitudinally by two husks when ripe.

STEM: Bark brown, grainy.

•On drying, the fruit splits open in two, exposing a single large chocolate-brown seed covered with a bright red, thick, branching, fleshy network (aril), like a coarse dishrag, which turns yellowish-brown and stiff on drying (illustrated). The seed itself is the nutmeg of commerce; the dried aril is mace. The use of nutmeg goes back to pre-history and it was one of the most valuable of spices in the Middle Ages.

Nerium oleander (syn. N. indicum)

Oleander
Rosebay

FAMILY: Apocynaceae

HABIT: Shrub, to 20 ft.

HABITAT: Ornamental; sometimes used as a hedge.

NATIVE TO: Mediterranean to Japan.

LEAF: Simple, opposite or in whorls of 3, stalkless or nearly so, stiff, narrow, thick, pointed, dark green above, lighter beneath; to 10 in. long, with a prominent midrib.

FLOWER: Showy terminal clusters; buds elongate, 5-parted, twisted, darker in color; petals wavy or crinkled, flaring to 2 in., yellow, rose, pink, purple, or white.

FRUIT: Elongate thin pods, typically in pairs, 4 to 7 in. long, green, then brown, drying and splitting lengthwise, with milkweed-like silky seeds within.

STEM: The branching pattern is primarily erect, usually with several stems in a cluster from the ground.

•Some oleanders are heavily scented, but others are entirely odorless. The flower has been doubled in some varieties (one illustrated). Unlike many of its relatives, oleander has no milky sap. It is very poisonous, however, containing principles that affect the heart. Meat skewered on oleander branches for cooking, can become lethal. Oleanders are sometimes found as large pot plants in the north and are common outdoor landscape plants in Florida and California. They are drought resistant and adapt well to places where care will be minimal. See also yellow oleander (Thevetia neriifolia).

Ochna serratula

Birdseye bush
Ochna
Mickey-mouse plant

FAMILY: Ochnaceae

HABIT: Shrub, to 8 ft.

HABITAT: Ornamental.

NATIVE TO: South Africa.

LEAF: Simple, alternate, almost stalkless, blade elliptical, pinnately veined, to 2½ in. long; edges sawtoothed.

FLOWER: Single or widely-spaced clusters. The 5 evenly spreading delicate yellow petals, about 1 in. across, are short-lived. The persistent calyx of 5 lobes, green at first, bends almost inside out and turns brilliant red. The stem beneath also thickens enormously and turns the same red.

FRUIT: One to several nearly spherical berries, green at first, then glossy black, from the face of the red, pad-like stem which also bears the persistent sepals. Each berry contains a single large seed.

STEM: Roughened with scars.

•The unusual, brilliantly colored and long-lasting sepals, associated with the conspicuous fruit and swollen red stem, give ochnas a dramatic, eye-catching appearance when in fruit.

Ophiopogon japonicus

Mondo grass
Dwarf lilyturf

FAMILY: Liliaceae

HABIT: Sod-forming herb.

HABITAT: Cultivated for lawns, slopes, etc.

NATIVE TO: Japan and Korea.

LEAF: Numerous, thickly clustered, very narrow, stiffish, each about 1/8 in. wide with 5 to 7 parallel veins lengthwise; erect but curving over; dark green, to 15 in. long; not deciduous.

FLOWER: Few-flowered erect clusters; flower small, nodding, pale lilac or white, 6-parted, lily-like.

FRUIT: A bluish berry.

ROOT: Long underground creepers, with tuberous enlargements.

•This plant is cultivated primarily for turf or garden edging. It is not mowed and stands shade and dryness well, giving a dense, shaggy, dark green low-maintenance cover in difficult locations. One variety has variegated foliage. A lawn of mondo grass is often recognized at a distance because of its pattern of swirls and "cowlicks" that develop as patches of "grass" all lean over in one direction or another.

Opuntia repens
Opuntia rubescens

Prickly pear
Tuna

FAMILY: Cactaceae

HABIT: O. repens – sprawling, to 2 ft. high, O. rubescens – erect, to 20 ft.

HABITAT: Weedy perennial invaders of dry lands.

NATIVE TO: West Indies.

LEAF: Small, inconspicuous; soon deciduous.

FLOWER: Single, almost stemless, growing close to the green cactus surface; yellow fading to reddish, 1 1/2 in. in O. repens; or yellow to red, and less than 1 in. in O. rubescens; delicate and short-lived in both.

FRUIT: Spherical, red; about 1 in. in O. repens, twice that in O. rubescens.

STEM: Flattened green segments, jointed in series or branching at joints; with evenly spaced clusters of spines, and often with fine wool, on all surfaces.

•Opuntias grow wild from Massachusetts to the Straits of Magellan. Most of the 300-odd species, like these two, are readily recognized by the flattened, platter-like stem segments, each less than 1 ft. long, usually distinctive and conspicuous in the vegetation wherever they grow. The fleshy stem segments conserve moisture, yet being flattened, take good photosynthetic advantage of available light.

Orchids

Orchids

FAMILY: Orchidaceae

•The tropics worldwide are rich in orchids. Authorities have estimated that there may be as many as 30,000 species of orchids altogether and additional thousands of named hybrids among some of these species are in cultivation.

Orchids represent advanced evolution among plants. They are specialized in a number of ways, especially in reproductive mechanisms, and their seed is almost microscopic. Many are showy and choice and have been brought into cultivation as house plants. Others are less conspicuous but are occasionally noticed in the wild flora.

While some orchids live in soil, most tropical orchids are epiphytes, growing among the branches or lodged on the trunks of trees of the rain forest (see further definition at Ficus spp.). Although attached to a host by their roots, they are not at all parasitic on it, depending entirely on their own vegetation to provide their sustenance through photosynthesis, and on their roots for absorbing minerals and water only from the surface of the host tree. The leaves, also, serve to catch rainwater, and may be modified in some species (swollen) to store water.

The orchid pictured here is an epiphyte native to the West Indies. Persons wishing to identify wild or cultivated orchids or learn more about them should consult the various specialized guides to orchids that are available.

Pandanus odoratissimus
Pandanus utilis

Screw pine
Walking pine
Hala

FAMILY: Pandanaceae

HABIT: Shrub or much branched tree, some to 60 ft.

HABITAT: Ornamental; naturalized.

NATIVE TO: Western Pacific and Indian Ocean.

LEAF: Clustered near branch tips, simple, strap-like, stiff, bent or arched, stalkless, with small spines along the edges, keeled at attachment; 4 in. by 6 ft.; leaving persistent leaf scars on the trunks after leaf drop. Some varieties have variegated foliage (illustrated).

FLOWER: Male and female flowers are on separate trees. Females occur in spherical clusters, males in drooping brush-like whitish clusters (illustrated); individual flowers of both sexes small, inconspicuous.

FRUIT: Large, fleshy, hanging, pineapple-like in size and appearance, green then usually orange. The units (carpels) are free or nearly so (unlike pineapple).

STEM: Conspicuous straight prop roots extend mostly at an angle from the trunk into the ground. These could be confused with the aerial roots of some figs or mangroves, but the leaves are entirely different.

•Screw pines are most easily recognized by the prop roots. Species differ in such characteristics as height and branching of the trunk, and the spininess of the leaf edges. Horticultural and botanical names of the various screw pines are not always clearly established as yet.

On another level, the plants themselves are often casually confused with palms by some and pineapples by others. They are related to neither. The "pine" in their name comes from the common name for pineapple in the tropics, given originally because the fruits look something like pinecones. "Screw" comes from the spiral pattern made by the leaf scars in young trees like the threads of a screw. The individual dried seedpods (illustrated) have a tufted pointed tip, like a paintbrush, and were used by native Pacific peoples for painting tapa cloth and themselves. The leaves are used for baskets, mats, thatching, and the like.

Parkinsonia aculeata

Jerusalem thorn

FAMILY: Leguminosae, Caesalpinioideae

HABIT: Tree, to 30 ft.

HABITAT: Ornamental; naturalized.

NATIVE TO: Tropical America.

LEAF: Alternate, 2x pinnate; leaflets in 10 to 25 pairs, minute, quickly shed; leaving 1 to 2 pairs of naked, flattened, green, photosynthetic, secondary leaf stalks arising from a short, primary leaf stalk. These are narrow, coarsely needle-like, 1 to 1½ ft. long, with parallel edges scarred by the points where the short-lived leaflets were attached.

FLOWER: Clustered; small, flat-open flowers, about 1 in. across, bright lemon yellow, each often with 1 reddish petal; fragrant.

FRUIT: Cylindrical woody brown pods, coarsely lumpy, 2 to 6 in. long.

STEM: Open airy tree with drooping branches.

•Scattered spines, to 1 in. long, represent persistent stipules (leaf structures that grow at or near the attachment of the leaf stalk to the stem bearing it). The general aspect of this tree is not unlike Casuarina, for the same leafless reason. Its "needles," however, are composed entirely of flattened leaf stalks with the leaflets gone, and not of delicate stems as in Casuarina. The "needles" of a Jerusalem thorn are usually brighter green than those of Casuarina. Before they open and are lost, the short-lived leaflets of Jerusalem thorn are folded back obliquely across the widened leaf stalk bearing them, their length barely exceeding its width (see illustration).

Passiflora spp.

Passion flower
Granadina

FAMILY: Passifloraceae

HABIT: Vines with tendrils.

HABITAT: Ornamental and naturalized.

NATIVE TO: Tropical America.

LEAF: Simple, alternate, stalked; blades smooth-edged or 3 to 5 lobed.

FLOWER: Red or white or blue or purple depending on species; the individual flower is large, showy, radially symmetrical, and complex. A whorl of about 12 spreading or reflexed petals surrounds 1 or more whorls of narrow tubular structures (not sterile stamens despite their appearance), with a fusion structure of stamens and style at the center.

FRUIT: Firm-walled, many-seeded, elongate or spherical depending on species, large, 3-sectioned, with numerous seeds.

•The complex structure at the center of a passion flower symbolized the crucifixion to early Spanish explorers.

The fruit is edible, the pulp eaten directly or used for juice. There are some 400 species of passionflower, two or three of which are grown commercially for the juice. Some of the showy species make spectacular pot plants when they flower. Some species produce large tubers from which the vines grow. The general description above applies to most of the species of Passiflora, but a few may vary in some details.

Pedilanthus tithymaloides

Christmas candle
Slipper flower
Japanese poinsettia
Slipper spurge
Redbird

FAMILY: Euphorbiaceae

HABIT: Shrub to 6 ft.

HABITAT: Ornamental; may be sheared as a hedge.

NATIVE TO: Tropical Americas.

LEAF: Simple, alternate, broadly elliptical, sharply pointed, thick, waxy; no leaf stalk; midrib somewhat winged below; to 4 in. long; may be all green or variegated with white; sometimes also with red.

FLOWER: Terminal clusters of pink or bright red "slippers", each pointed, to 3/4 in. long, in diverging pairs; short-lived stamens or style protrude briefly from the point (male and female flowers separate).

FRUIT: A longitudinally striped, ovoid, 3-parted pod, about 1/2 in. long.

STEM: Thick and fleshy in some varieties. The stem often bends sharply at the points where leaves are attached, giving the plant an unusual zig-zag appearance (illustrated). If the leaves are variegated, the stem is also variegated with white stripes.

•This plant, like many others in this family, contains a milky sap which can cause skin irritation or irritation of the digestive system if eaten. The variegated form is more common than the solid green form in many places. The colored "slippers" are not formed by petals or sepals as might first be thought, but by rolled bracts (modified leaves associated with flowering). Birds are attracted to the seeds.

Persea americana

Avocado
Alligator pear

FAMILY: Lauraceae

HABIT: Tree, 30 to 60 ft.

HABITAT: Cultivated.

NATIVE TO: Central America.

LEAF: Simple, alternate, crowded in whorls near branch tips; stalked, 4 to 8 in. long by about 3 in. wide; blades broadly elliptical, sharp-pointed, pinnately veined, shiny; not deciduous.

FLOWER: Inconspicuous, yellowish green, in clusters often of hundreds.

FRUIT: The familiar avocado of commerce. Size, shape, and color vary with horticultural variety. Some kinds are warty; others smooth-surfaced. Each contains a single large stone.

•Archeological evidence has shown that this fruit was cultivated as much as 8,000 years ago in Central America. The cultivars of avocado are now divided into three groups: Mexican, West Indian, and Guatemalan. Some are pear-shaped; others spherical.

Each avocado flower opens twice, first as a functional female, then about a day later as a male. This staggered timing assures cross pollination, but only one in several thousand flowers actually sets fruit. Avocado fruits are remarkable for their high content of oil which can be as much as 30 percent of the weight of the flesh.

Petrea volubilis

Sandpaper vine
Purple wreath
Queen's wreath

FAMILY: Verbenaceae

HABIT: Vine, to 35 ft.

HABITAT: Ornamental.

NATIVE TO: Mexico and Central America.

LEAF: Simple, opposite, short-stalked, elliptical, stiff, pointed, pinnately veined, very rough surfaced; 2 to 8 in. long.

FLOWER: Clustered, numcrous; each 5-parted, starlike; sepals long-lived, pale lavender, narrow, elongate; petals dark purple, wider, lasting only a day or two.

FRUIT: Inconspicuous.

STEM: Woody; often arching over and down.

•The flowers are clustered in single or grouped lilac-like cascades. When the short-lived petals are present, the flower looks something like an African violet. A white-flowered variety (illustrated) is occasionally encountered.

Sandpaper vine takes this name from the characteristic sandpapery nature of the leaves. This is usually a surprise when encountered since the leaves don't look particularly rough. The roughness and the unusual two-blue flower, together are an unmistakable identifier. Petrea is often grown as a hedge and may be clipped.

Phoradendron spp.

Mistletoe
False mistletoe
American mistletoe

FAMILY: Loranthaceae

HABIT: Shrubby, woody parasite.

HABITAT: Parasite, growing on the branches of other trees.

NATIVE TO: Tropical Americas.

LEAF: Simple, opposite, greenish (despite parasitic nature), leathery, nearly stalkless; the size and shape depend on species, but mostly small and apt to be wider beyond the middle.

FLOWER: Small, inconspicuous.

FRUIT: A small berry (color depends on species) with gluey pulp.

•The gluey nature of the fruit helps it adhere to the branches of a host where it germinates and takes root. The host shown in the photographs is Casuarina since the difference in foliage makes the Phoradendron stand out, but mistletoe grows equally well on a variety of broad-leaved trees where its own foliage is more hidden. The roots penetrate the bark of the host and draw sustenance from the host. Mistletoes are serious parasites of trees and shrubs in many tropical areas. They are not particularly choosy about host. A severe infestation of mistletoe will eventually kill the host tree, even large ones. When it does this, the mistletoe also defeats its own purposes, since without sustenance, it too will die.

The mistletoe of legends, lore, and literature is the related Old World Viscum album. The white-berried mistletoe sold in American shops at Christmastime is usually Phoradendron serotinum. Some 200 species of Phoradendron are distributed throughout the tropics.

Piper nigrum

Pepper
Black pepper
White pepper

FAMILY: Piperaceae

HABIT: Woody vine.

HABITAT: Cultivated on posts or living trees.

NATIVE TO: India and Ceylon.

LEAF: Simple, alternate, stalked, 3 to 6 in. long; blade oval, pointed, with conspicuous veins running mostly the length of the blade and appearing almost palmate.

FLOWER: Clustered in hanging spikes as long as the leaves; individual flowers inconspicuous, greenish, lacking petals or sepals.

FRUIT: Small, green then red berries, about 1/4 in. in diameter. Stages from inconspicuous flowering spikes to full-sized green berries are shown in the top right illustration.

•Dried unripe fruits make the black peppercorns of commerce with their thin wrinkled skins. Milder white pepper is made from fully ripe fruits from which the skin is removed in a fermentation process. These condiment peppers are not to be confused with the vegetable pepper, a species of Capsicum from an entirely different plant family. In many places, vegetable peppers are called bell peppers to distinguish them from the condiment kind. Bell peppers include chili pepper.

Pisonia subcordata

Water mampoo

FAMILY: Nyctaginaceae

HABIT: Tree, to 50 ft.

HABITAT: Deciduous woods; also planted as a shade tree.

NATIVE TO: West Indies.

LEAF: Simple, opposite, short-stalked, to 6 in. long or more; blade broadly elliptical, upper surface shiny, under surface dull; blade tips rounded or pointed.

FLOWER: In fragrant clusters at the branch tips; small, inconspicuous, tubular, whitish, 5-parted.

FRUIT: Elliptical, 1/3 in. long; with small longitudinal ridges; red then black, succulent, sticky.

STEM: Bark smooth and blue-gray, like a beech tree or lightly corrugated in older parts.

•This pleasant shade tree has a wide-spreading crown and may become massive in time. It is planted occasionally as a shade tree and can withstand salt spray near shores which makes it doubly useful. Unfortunately, it looks like ye olde standard tree, and has no special vegetative characteristic that makes recognition easy.

Plumbago auriculata (syn. P. capensis)

Plumbago
Leadwort
Cape leadwort

FAMILY: Plumbaginaceae

HABIT: Woody shrub, 2 to 3 ft. tall; with sprawling stems to 15 ft.

HABITAT: Ornamental.

NATIVE TO: South Africa.

LEAF: Simple, alternate, elliptical, to 2 in. long, white and scruffy beneath; with a wavy surface, sometimes wavy edge; short or no leafstalk; not deciduous.

FLOWER: Clustered, light blue and also a white variety (both illustrated); each flower 5-parted, long-tubular; petals flaring about 3/4 in. across, delicate, with a distinct midvein, dark blue in the blue variety. The calyx tube, less than half the length of the petals it surrounds, bears conspicuous large, complex, sticky hairs.

FRUIT: A tiny dry pod, splitting into 5 segments, each 1-seeded. Seed pod dispersal is helped by the sticky persistent calyx hairs.

STEM: Slender, semi-climbing, long-arching.

•This plant is often pruned vigorously to reduce its inherant sprawl, and grown as a colorful hedge. Its particular shade of blue varies in intensity among different plants but is among the choicer blues of the garden. Another species (P. indica) has flowers that are similar in structure, but bright red. The common name comes from the lead-colored roots, a relationship reflected also in the genus name which is Latin for "lead".

Plumeria alba
Plumeria obtusa
Plumeria rubra

Frangipani
Plumeria

FAMILY: Apocynaceae

HABIT: Coarse shrubs or small trees, 25 to 50 ft. depending on species.

HABITAT: Ornamental.

NATIVE TO: Central America, Caribbean.

LEAF: Simple, alternate, narrow to broadly elliptical; smooth edges. The three species differ in other leaf details; see below.

FLOWER: Single or few-clustered; large, heavy, 5-parted; bud tubular and twisted; opening into flaring petals which overlap regularly at the base, one side over, the other under the next (imbricate).

FRUIT: Paired pencil-like leathery pods, 1/2 ft. or more long. Seeds with an excentric basal wing.

STEM: Stout, weak, blunt tipped, coarsely bumpy; illustrated lower right and in the caterpillar photograph.

•The basic branching pattern of a frangipani is into two equal branches at each branching point, like 'Y's. The three species can be told apart by the leaves alone: P. alba (native to the West Indies) – leaves thin, to 15 in. long, narrow, with somewhat rolled margins, hairy undersurface, no connecting vein, deciduous; P. obtusa – leaves thick, broader, 6 to 8 in. long, flat, smooth underneath, with a connecting vein along the edges (see illustration), not deciduous; P. rubra (commonest) – leaves as in P. obtusata but to 18 in. long, deciduous. In some places, plumerias are regularly seasonally defoliated by hords of a particular caterpillar (illustrated). Flowering soon follows defoliation.

The flower colors are as follows: P.alba and P.obtusata – white with a diffuse yellow center; P.rubra – red, yellow, orange, or white, often with a diffuse yellow center.

Podocarpus macrophyllus

Podocarp
Southern yew
Buddhist pine
Japanese yew

FAMILY: Podocarpaceae

HABIT: Tree to 35 ft.

HABITAT: Ornamental.

NATIVE TO: Southern Hemisphere and Japan.

LEAF: Simple, flat, narrow-elliptical, 3 to 4 in. long by 3/8 in. broad, with nearly parallel edges; glossy, dark green above, paler below; conspicuous midvein running the length; spiral placement on the stem, but often clustered more intensely at the branch tips; not deciduous.

FLOWER: Both sexes inconspicuous; male: small catkin-like clusters; female: small inconspicuous cones of 2 to 4 scales only, usually single-seeded.

FRUIT: The naked seeds of the female cone are berry-like, fleshy about 1/2 in. long, and borne on a fleshy pad of tissue or thickened stalk of about the same size.

STEM: The branching pattern is usually obviously horizontal.

•Podocarps are true conifers (gymnosperms), but tend in most species to have blade-like leaves rather than needles. This particular species has especially large leaves and is available in many cultivars varying in details of branching, leaf size and shape. It is a common doorway pot plant in northern hotels, withstanding cold drafts well. In the tropics, it can be clipped and makes an excellent dense hedge.

Podraena ricasoliana (syn. Pandorea ricasoliana)

Pink trumpet vine

FAMILY: Bignoniaceae

HABIT: Vine, often trellised.

HABITAT: Ornamental.

NATIVE TO: South Africa.

LEAF: Opposite, 1x odd-pinnate; leaflets 5 to 11, oval, about 2 in. long, coarsely toothed, the terminal leaflet often somewhat larger than the others.

FLOWER: Loosely clustered, showy, tubular flowers; each about 2 to 3 in. long, flaring into 5 unequal, crinkled lobes, the larger downward; petals light purple or pink, marked with dark purple or dull red longitudinal lines; the petal tube growing from an inflated "pocket" of joined sepals, white in color.

FRUIT: Thin, cylindrical leathery pod, to 15 in. long.

•The name is an anagram of Pandorea, its former scientific name, and to which plant as now understood, it is closely related. The plant is quick-growing, showy, and tolerant of poor soil and full sun. It is commonly planted to cover gazebos, fences, and the like.

Botanists continually study and learn more about plants. Sometimes newly discovered facts require that the relationships of the plant be reconsidered. This may result, as here, in the removal of a species from the genus in which it earlier was placed. This, in turn, requires the assignment of a new name. The old scientific names are hard enough to learn, and no one likes having to learn new combinations. But progress will out, like it or not.

Polyscias guilfoylei

Aralia
Geranium-leafed aralia

FAMILY: Araliaceae

HABIT: Shrub, to 20 ft.

HABITAT: Ornamental; often used as a hedge.

NATIVE TO: Polynesia.

LEAF: Alternate, basically single, simple, and pinnately divided, but further complicated by irregular incutting and coarse toothing of the edges; blades to 16 in. long; "leaflets" white-edged or variegated.

FLOWER: Very tiny, inconspicuous; flowering infrequent.

FRUIT: Small inconspicuous berries.

•Commonly planted as a hedge or screen that will tolerate difficult conditions yet grow readily and rapidly. Its basic open growth can be made denser by frequent clipping. The foliage is quite variable among the common cultivated varieties. The leaves are aromatic.

Although Polyscias is closely related to the genus Aralia, the two are not the same, despite the possible confusion caused by the common name of Polyscias.

Pseuderanthemum reticulatum (syn. Eranthemum reticulatum)

Golden eranthemum
Eranthemum
Yellow-vein bush

FAMILY: Acanthaceae

HABIT: Shrub, 3 to 5 ft.

HABITAT: Ornamental.

NATIVE TO: Polynesia.

LEAF: Simple, opposite, stiff, broadly oval, short-stalked, bluntly pointed; to 10 in. long. Young leaves are yellow, older becoming green, with a pinnate network of conspicuous yellow veining.

FLOWER: An upright flowering stalk, maturing base up, bearing 5 to 15 open flowers at any time. Each flower short-stalked, about 1 1/2 in. across, tubular, flaring into 5 petals, the two upper petals usually more closely associated than the three lower and sometimes appearing as one; throat and inner parts of petals speckled or dotted with deep wine-purple.

•Actively growing plants are usually conspicuous as a sunny yellow foliage against a darker green background. The plant flowers nearly continuously in the tropics. Another species, P. atropurpureum, has leaves that are splotched with light or dark purple and flowers conspicuously marked with purple.

Psidium guajava

Guava

FAMILY: Myrtaceae

HABIT: Shrub or tree, to 30 ft.

HABITAT: Cultivated and widely naturalized or wild.

NATIVE TO: Tropical America.

LEAF: Simple, opposite, short-stalked, blade broadly elliptical, smooth-edged, to 6 in. long; hairy beneath; veins prominent; not deciduous.

FLOWER: Solitary or few; white, radially symmetrical, to 3 in. across; a pincushion of numerous protruding stamens inside a whorl of 5 spreading petals.

FRUIT: Ovoid, 1 to 4 in. long, thin-skinned, green or yellowish, tipped with a persistent calyx; flesh yellow or dark pink, aromatic, sweet, used in jellies. Seeds numerous, hard.

STEM: Trunk brown, bark scaly. Gray-green branchlets, somewhat 4-angled.

•The seeds are scattered by birds. This plant, weedy in some areas, is found world-wide, but is especially important in India. It begins bearing at about two years, reaching maximum in about five more, continuing for thirty years. The fruits are in increasing demand for the juice which is rich in vitamin C. A seedless variety has been developed. Externally, the fruits can be confused with pomegranite (Punica) because of the persistent calyx.

Punica granatum

Pomegranite

FAMILY: Punicaceae

HABIT: Shrub or small tree, to 20 ft.

HABITAT: Cultivated for fruit and as an ornamental.

NATIVE TO: Persia.

LEAF: Simple, opposite, short-stalked, broader outward, smooth-edged, glossy; to 3 in. long.

FLOWER: Solitary or few at branch tips; showy, tubular; calyx tube 5- to 7-pointed, yellowish, orange or purple, persistent; petals crepe-like, orange-red, opening weakly to 1½ in. across; stamens numerous within; also doubled flowers and other colors in horticultural varieties.

FRUIT: Subglobose, to 5 in., blushing yellow-brown-red when ripe; seeds numerous in a juicy red-purple pulp; calyx persistent at fruit tip.

STEM: Branches are occasionally tipped with spines.

•Pomegranite has been a major fruit of the Mediterranean region since prebiblical times. Individual trees may live a century or more. Selected varieties have been developed and dwarfed for ornamental use, with white, red, mixed, and double-flowered kinds. It is sometimes grown as a hedge. With care, if the ripe fruit is squeezed or rolled between the hands, the slightly acid juice can be sucked out of it with a straw. This avoids the problem of the numerous seeds.

Ravenala madagascariensis

Travelers tree
Travelers palm

FAMILY: Strelitziaceae

HABIT: Herbaceous tree.

HABITAT: Naturalized; occasionally planted as an oddity.

NATIVE TO: Madagascar.

LEAF: To 20 ft. long; leaf stalks slightly longer than blades; leaf bases hollow, folded, and clasping; young leaves with a large, oval, smooth-edged blade; older blades torn along veins and thus palm-like.

FLOWER: Clusters of about a dozen floral units eventually arise between the leaf bases, each resembling a large compound bird-of-paradise (Strelitzia), and composed of a large, long, boat-shaped green bract with many whitish flowers within.

FRUIT: Look like small woody bananas with steely blue seeds within.

STEM: Tipped with banana-like leaves which arise two-ranked in a vertical plane, bases overlapping, giving the appearance of a large symmetrical fan at the top of the stem.

•Young plants start with a fan of leaves arising at ground level. Unlike its relative the banana (Musa), the main bud slowly creates a solid woody stem (trunk) which can attain 30 or 40 ft. or more in height, carrying the typical fan of leaves at the top. This plant lives for many years. It is called traveler's palm because the leaf bases store up to a quart of potable water (if you ignore occasional drowned insects), and it looks somewhat like a palm (but it is not a true palm).

A young Ravenala can be confused with Strelitzia nicholai. They can be separated by the bracts which are green in Ravenala as opposed to colored in both species of Strelitzia, and by the trunk which is solid and woody in Ravenala as opposed to hollow in Strelitiza.

Rhizophora mangle

Red mangrove
American mangrove

FAMILY: Rhizophoraceae

HABIT: Shrub or tree, to 40 ft.

HABITAT: Edges of shallow marine embayments.

NATIVE TO: New World.

LEAF: Simple, opposite, thick, leathery, elliptical; stalked, to 6 in. long; secondary veins perpendicular to the main vein and regularly parallel to each other.

FLOWER: Regular, 4-parted; about 3/4 in. across; petals spreading, stiffish, yellow, hairy.

FRUIT: Ovoid, about 1 1/4 in. long, germinating in situ to form an elongate root, to 1 ft. long (illustrated), with the beginnings of leaves at the top, before dropping.

STEM: Gray bark, reddish underneath. Prop roots (illustrated) arise from the trunks and grow downwards through the seawater to the bottom.

•Mangroves tolerate salt water and are able to colonize shallow marine waters. Red mangrove, which is able to grow the most seaward of all, uses its prop roots to "march out" into the water, forming dense tangles of vegetation and trapping sediment from the passing water currents. This, in turn, provides a habitat for a rich assemblage of other organisms. It propagates also by the precocious seeds which germinate in place on the branches and are well adapted to plunge upright into the water from overhanging branches and embed themselves in the mud.

Other mangroves lack conspicuous prop roots. The black mangrove (Avicennia nitida) usually grows just landward of the red mangrove and is conspicuous by its pneumatophores. These are coarse, erect, pencil-like sticks arising out of the water from the roots. White mangrove (Laguncularia racemosa) and buttonwood (Conocarpus erecta), illustrated upper center, tolerate only occasional flooding by seawater. A variety of buttonwood with densely hairy whitish leaves, illustrated upper right, is occasionally planted as an ornamental. Twigs illustrated lower right are of Avicinnia nitida, Laguncularia racemosa, and Conocarpus erectus from left to right.

Rhoeo spathacea (syn. R. discolor)

Moses-in-a-cradle
Purple-leafed spiderwort
Boat lily
Oyster plant

FAMILY: Commelinaceae

HABIT: Perennial herb.

HABITAT: Ornamental.

NATIVE TO: Tropical America.

LEAF: Clustered near the ground at the tip of a short stem; mostly erect, narrow, pointed, stalkless, clasping, with parallel veins; about 1 ft. long by 3 in. wide; dark green above (inward), rich purple under (outward).

FLOWER: At base of leaves, enclosed in 2 boat-shaped bracts (small modified leaves), the white flowers are barely visible within; flowers 3-parted, the white petals soon disappearing.

FRUIT: Small spherical pods hidden within the persistent bracts.

•This plant is grown primarily for its colorful foliage, and is a familiar, easily cultivated pot plant in the north. The flowers and fruits hidden in their "boats" (or "cradles") intrigue the imagination of children. A variegated variety has pale yellow stripes down the leaves. Rhoeo spathacea is readily recognized by its foliage and the unusual inflorescence, though it can be confused superficially with some other plants with deep purple foliage.

Ricinus communis

Castor bean
Palma christi

FAMILY: Euphorbiaceae

HABIT: Rank herb to 15 ft. or more.

HABITAT: Ornamental; cultivated crop; naturalized.

NATIVE TO: Tropical Africa.

LEAF: Simple, alternate, long-stalked; blade 1 to 3 ft. across, palmately cut into 6 to 11 lobes with toothed edges; leaf stalk attached slightly inside the blade edge, with main veins radiating from that point.

FLOWER: Sexes separate, but in the same spiny spike; inconspicuous females above, males with white anthers below, both without petals.

FRUIT: Spherical or slightly elongate pods with fleshy spines; drying and opening into 3 chambers, 1 large seed (1/2 in. long or more) in each.

•Castor bean is cultivated as a crop for the oil which can be pressed from the seeds. Castor oil has minor use in medicine, but major use as a machine lubricant. The plant is also grown as a quick screen or coarse hedge. The plant is variable in size, and in the degree of red pigmentation in the foliage and stems (see illustrations). Fruits may be brilliant red in some; green in others. Seeds are severely poisonous, yet are readily available in seed packets in supermarket seed displays, etc. They are sometimes strung in necklaces. The seeds, which look like large ticks, can remain dormant for long periods in the soil.

Roystonea regia

Royal palm
Cuban royal palm

FAMILY: Palmae

HABIT: Stately tree, to 75 ft.

HABITAT: Ornamental, especially lining drives and avenues; sometimes naturalized.

NATIVE TO: Cuba.

LEAF: Alternate, 1x pinnately compound palm leaf with sheathing base which with others forms a prominent, long, upward, smooth green tube at the top of the tree; blade to 15 ft. long; leaflets numerous, each leaflet about 3 ft. long by 2 in. wide.

FLOWER: Sexes separate but on the same tree; inflorescences, sometimes 3 to 4 ft. long, arise from the trunk just below the leaf sheaths; both male and female flowers small and inconspicuous.

FRUIT: Tiny oblong berries, initially green, then red, finally purple-black.

STEM: Trunk is typically straight, smooth, light gray; it may bulge slightly at the base and part way up. The surface is finely marked (see closeup). The horizontal lines represent the scars where the leaf-stalks were attached.

•This is a fast-growing palm. The trunk looks as though made of smooth cement. The genus is named after Roy Stone, an engineer important in the early development of Puerto Rico. Roystonea borinquana is closely related and similar. These are among the most striking and statuesque of all palms. Their fronds and fruits are used in native cultures in a variety of ways.

Russelia equisetiformis

Coral blow
Fountain bush
Firecracker

FAMILY: Scrophulariaceae

HABIT: Bush, to 4 ft.

HABITAT: Ornamental; naturalized.

NATIVE TO: Mexico.

LEAF: Small, simple, sparse (see closeup illustration), 3 to 6 in a whorl; elliptical, about 1/2 in. long or reduced to scales; early deciduous.

FLOWER: In loose clusters of 1 to 3; long-tubular, small, bright scarlet, hanging, flaring slightly into 5 lobes divided into 2 groups, upper and lower, with 4 stamens within.

STEM: Whorls of slim, smooth, arching, green, prominently-ridged branches which look something like horsetail stems (whence the species name).

•In an evolutionary sense, this plant is giving up on leaves for its basic photosynthetic function and transferring food-making increasingly to the branches which have become bright green and very numerous in consequence. This makes for an unusual soft-textured hedge or sprawling ground cover. The numerous bright red flowers look like hanging firecrackers and are particularly attractive to humming birds.

BOTANICAL GARDEN

Saccharum officinarum

Sugar cane

FAMILY: Gramineae

HABIT: Perennial grass, 8 to 20 ft.

HABITAT: Cultivated.

NATIVE TO: Tropical southeast Asia.

LEAF: Simple, alternate, numerous, grasslike, erect, in 2 rows along the stem; parallel-veined, 2½ in. wide, clasping the stem at base. Older leaves die and turn brown but do not fall from the stem.

FLOWER: Upright brushy clusters, to 2 ft. tall at the top of the stem, freely branched, feathery. Individual flowers small and inconspicuous.

FRUIT: A grain, but not used as seed or in any way commercially.

STEM: Erect, to 2 in. thick, solid, juicy, leafy the whole length; suckers freely at the base and thus makes clumps of stems; internodes are 6 to 10 in. long, each with a single strong lateral bud (see illustration).

•From southeast Asia, sugar cane reached India by the 3rd century BC, Egypt by the 6th AD, Spain by the 7th, the West Indies by the 15th, and Hawaii early in the 19th. The stem contains about 80 percent juice, and sucrose is 14 to 20 percent of that. The plant is harvested at flowering. Sugarcane is propagated by cane cuttings year round. Time until harvest is 10 to 20 months depending on variety and location. The greatest sugar concentration in the stem is near the ground. Thus stems must be cut at ground level, definitely stoop labor.

Samanea saman

Saman
Raintree
Monkey pod tree

FAMILY: Leguminosae, Mimosoideae

HABIT: Massively spreading short-trunked tree, to 80 ft.

HABITAT: Naturalized; shade tree.

NATIVE TO: Central America.

LEAF: Alternate, 2x odd-pinnate; 2 to 6 pairs of primary units each with 8 pairs of leaflets; terminal leaflet larger, to 2½ in. long; dry-season deciduous.

FLOWER: Pink feathery tufts, spring to fall; individual flowers small, inconspicuous, except for the numerous protruding stamens.

FRUIT: Flat legume pod with a characteristic slightly raised border down both sides (illustrated), green then black, to 8 in. long.

STEM: Fast growing; the tree spreads wider than tall, the top taking the shape of a shallow umbrella.

•This species was introduced to plantations as a shade tree. The young pods make good forage, and the fine-grained, striped wood is used for furniture and ornaments. Pasture or lawn under a saman is usually greener than elsewhere. Some say this is so because the leaflets close up at night and in rainy times let the moisture through.

Sansevieria trifasciata

Snake plant
Mother-in-law's-tongue
Bowstring hemp

FAMILY: Agavaceae

HABIT: Rhizomatous perennial herb.

HABITAT: Ornamental; naturalized.

NATIVE TO: Africa and Southeast Asia.

LEAF: 1 to a few; erect, stiff, stalkless, strap-like, thick, pointed; to 4 ft. tall by almost 3 in. wide in the common kind; variegated in cross bands or splotches, markings mostly vertical in some or along edges.

FLOWER: A single erect inflorescence; flowers numerous, narrow, elongate, tubular, white or off-white, 6-parted, fragrant; radiating horizontally on very short stalks from the main axis.

STEM: No visible stem; leaves grow from the rhizome (underground stem) in a clump or sometimes singly at ground level.

•This species has many cultivars. Some are short and squat (see illustration) and look as though they should be a different species. Sansevieria naturalizes easily and has become a major roadside weed in some parts of its range. Under such conditions it usually looks shabby and well-worn, but cared for, it can make a handsome specimen and is a common houseplant in the north.

Scaevola frutescens

Beach naupaka
Scaevola

FAMILY: Goodeniaceae

HABIT: Large shrub, to 10 ft.

HABITAT: Ornamental; occasionally as a hedge.

NATIVE TO: Pacific and Indian Ocean islands.

LEAF: Simple, clustered, almost stalkless, mostly held erect; blunt-tipped, smooth-edged, broader beyond the middle, bright green on top, thick, glossy, to 6 in. long; midvein distinct. Some may be slightly toothed along the outer edges.

FLOWER: Clustered close to the leaf bases, each white with radiating purple lines and blush, asymmetrical, with 5 to 6 petal lobes, each about 3/4 in. long, radiating to the sides and downward.

FRUIT: A fleshy white berry, about 1/2 in. in diameter.

•The flower bud is tubular, but splits along the top surface lengthwise on opening so that all the petal lobes come to open mostly downward. Each flower, thus, looks like an incomplete half-flower. Its one-sided flowers, though small, attract attention as a curiosity and help identify it easily. In one variety of this species, the leaves are covered with fine, dense down.

This handsome shrub endures salt spray and wind well and is increasingly found in seaside gardens or landscaping throughout the tropics.

Schinus terebenthifolius

Brazilian pepper tree
Florida holly
Christmasberry tree
Peppertree

FAMILY: Anacardiaceae

HABIT: Shrub to tree, to 40 ft. or more.

HABITAT: Ornamental; naturalized, weedy potential.

NATIVE TO: Brazil.

LEAF: Alternate, 1x odd-pinnate; 5 to 11 leaflets on a usually reddish stalk; each broadly elliptical, stalkless, with a dark green upper surface; to 3 in. long; veins pinnate, conspicuous, sometimes reddish.

FLOWER: Clustered; tiny yellowish-white, inconspicuous (illustrated upper right); male and female on separate plants.

FRUIT: On female plants; abundant small berries in loose hanging clusters; each berry 3/16 in., green then red, aromatic, juicy, resinous, peppery to taste.

STEM: Rangy-branching. Old tree trunks become gnarled and twisted.

•The berries are easily spread. Each contains a single yellow kidney-shaped seed which germinates readily. Pepper tree is an aggressive, troublesome, weedy species in many places. The Peruvian pepper tree (Schinus molle) is similar, but with smaller, narrower, and more numerous leaflets. The fruiting sprays of both species are often gathered for Christmas ornamentation. The berries can cause indigestion.

Senecio confusus

Mexican flame vine
Orange glow vine

FAMILY: Compositae

HABIT: Vine or sprawling shrub, to 25 ft.

HABITAT: Ornamental.

NATIVE TO: Mexico.

LEAF: Simple, alternate, elongate-heart-shaped, short-stalked; blades blunt or slightly heart-shaped at attachment to stalk, pointed, coarsely toothed, leathery, to 4 in. long.

FLOWER: Clusters of small daisy-like heads, with radiating bright orange rays ("petals") surrounding a loose mound of florets in the center (see illustration).

STEM: Sprawling, woody, rooting at nodes.

•This is a fast growing plant well suited to dry locations and full sun. It can be managed as a ground cover or as a vine. The rays of the composite head tend to bend backward and are narrower than those of Wedelia trilobata with which it might be confused. The leaves serve to separate these plants easily. Those of Senecio are alternate while those of Wedelia are opposite.

Sesuvium portulacastrum

Beach purslane
Sea purslane

FAMILY: Polygonaceae

HABIT: Sprawling herb.

HABITAT: Beaches and salt marshes, essentially at sea level.

NATIVE TO: West Indies.

LEAF: Simple, opposite, narrow, swollen-fleshy, but slightly or moderately flattened, to 2 in. long; reddish green or green.

FLOWER: Inconspicuous pink or white, in leaf axils; 5 radiating petal lobes around a green center, about 1 in. across; petals end in a small spine-like projection.

FRUIT: Small, containing tiny jet black seeds.

STEM: Fleshy, rooting at the nodes. Often yellowish in older parts of the plant.

•This plant creeps over beach sand with its long runners. It can tolerate full-salinity sea water and is now found on beaches in the tropics around the world. Sesuvium maritima is similar but the leaves are an inch or less in length and somewhat broader. Both species live in the same kind of beach habitat as bay bean (Canavalia maritima) and seaside morning glory (Ipomoea pes-caprae). The somewhat sausage-shaped leaves immediately separate Sesuvium from the others. All of these plants are important beach stabilizers.

Setcreasea pallida (syn. S. purpurea)

Purple heart
Purple queen

FAMILY: Commelinaceae

HABIT: Sprawling herb, to about 1 ft. high.

HABITAT: Ornamental.

NATIVE TO: Mexico and south Texas.

LEAF: Simple, alternate, deep purple both surfaces, clasping the stem, narrow, pointed, parallel-veined, to 7 in. long by 1 in. wide; trough-shaped especially when young.

FLOWER: Small, mostly at stem tips, short-stalked, 3-parted; petals pale to medium pink or purple, about 3/4 in. across.

STEM: Purple, somewhat fleshy, mostly creeping, readily rooting.

•This is one of several green, purple, or variegated plants of similar habit that serve as common pot ornamentals in the north. In the tropics they live outdoors in gardens as edgings for walks, requiring little care. It can be distinguished from Rhoeo spathacea by its leaves which are much larger and entirely purple instead of having one green surface.

Solandra maxima (syn S. nitida)

Chalice vine
Cup of gold

FAMILY: Solanaceae

HABIT: Woody vine climbing to 50 ft. or sprawling shrub.

HABITAT: Ornamental.

NATIVE TO: Mexico.

LEAF: Simple, alternate, short-stalked, to 6 in. long; blade oval, some broader beyond the middle, pointed, leathery, somewhat thick, glossy, pinnately veined.

FLOWER: Huge, solitary, tubular, 5-parted, to 9 in. long; corolla tube bulging, with petal tips bent back; opening off-white, yellowing with age, marked with longitudinal purple lines; fragrant.

FRUIT: Spherical, 2½ in. in diameter, forming within the persistent sepals; white or cream, edible, mildly acid.

STEM: Viney or sprawling; may possess or lack aerial roots.

•This is a conspicuous winter-blooming plant. Its gigantic flowers invariably attract particular attention. Each flower is individually spectacular. When flowering, the long waxy buds open with incredible speed, the motion of the petal lobes being just visible to the eye. The flowers last about four days each.

The plant contains an alkaloid and the sap is dangerous if gotten in the eye.

Spathiphyllum clevelandii

Spathiphyllum
White anthurium

FAMILY: Araceae

HABIT: Perennial herb.

HABITAT: Ornamental.

NATIVE TO: Tropical America; this is a cultivar of uncertain exact origin.

LEAF: Clustered at the ground on short stems; each leaf long-stalked; stalk sheathing; blade elliptical, pointed at both ends, to 1 ft. long and 2½ in. wide, dark green, glossy on the upper surface, pinnately veined, edges more or less wavy or slightly incut.

FLOWER: Single, erected above the foliage; composed of a spathe and spadix (see Anthurium). The spathe is white, erect, pointed, clasping the spadix stem at its base, thin delicate but somewhat waxy in texture, to 6 in. long; the columnar spadix is white, distinctly shorter than the spathe, bearing numerous, small, bisexual flowers.

FRUIT: Small scattered berries.

•A number of species of this genus, differing in size and detail of floral structure, are in the horticultural trade. Many hybrids exist, which makes exact identification difficult. Smaller ones make excellent pot plants. The inflorescence is long-lasting like that of Anthurium to which it is closely related.

Spathodea campanulata

African tulip tree
Flame of the forest

FAMILY: Bignoniaceae

HABIT: Tree, to 60 ft.

HABITAT: Ornamental.

NATIVE TO: West Africa.

LEAF: Opposite or in whorls of 3, 1x odd-pinnate, to 18 in. long; leaflets 4 to 8 pairs, broadly elliptical, smooth-edged, glossy, pointed, pinnately veined, almost stalkless, 3 to 4 in. long; not deciduous.

FLOWER: Clustered in a round clump of curving erect, swollen, brownish green buds, outer flowers opening first; flowers bright orange-red, often with yellow overtones and dark lines; upturned, somewhat tubular, but bilaterally symmetrical, 3 in. across by 5 long. The tubular petals form a deep-bellied cup edged with 5 shallow frilly petal lobes. The calyx is leathery.

FRUIT: Erect brown boat-shaped pod, 8 to 10 in. long, by 2 in. in diameter; containing hundreds of shiny seeds with papery membranes.

STEM: An erect tree with upward branching pattern. The flower clusters are scattered at the ends of the branches in the tree tops (see illustration upper left).

•The tree has religious significance in some areas of Africa. The flower buds become filled with water as they develop which is an unusual feature among plants. In a yellow-flowered cultivar, the flowers are entirely bright lemon yellow. The yellow-flowered variety with fruiting pods is illustrated in the center left.

Stapelia gigantea

Stapelia
Carrion flower
Starfish flower

FAMILY: Asclepiadaceae

HABIT: Low-growing succulent.

HABITAT: Ornamental.

NATIVE TO: Africa.

LEAF: None.

FLOWER: Huge, mostly single, borne on a short stalk arising near the base of the plant; 5 partly separate long-triangular petal lobes from a funnel throat, leathery, dull yellow, purple, or red-brown, to 16 in. wide; petals marked with darker transverse lines or splotches, opening from a conspicuous angularly swollen and pointed bud.

FRUIT: Pods usually borne in pairs (illustrated in inset), containing seeds tufted with silky hairs as in common milkweed.

STEM: Clusters of erect, angled, gray-green, fleshy stems, about 1 in. thick or a little more by 6 to 8 in. tall and cactus-like; ridges are toothed with points but lack real spines.

•The flower has a strong unpleasant carrion odor and is pollinated by the kind of insects attracted to rotting meat. The scientific name can be confused with Styphelia, but the plant can hardly be confused with anything else when in flower. The pods and silky seeds give a good clue to the milkweed relationship of this plant which is otherwise more cactus-like in general aspect. Unlike most cacti, it has no spines.

Stephanotis floribunda

Stephanotis
Bridal bouquet
Madagascar jasmine

FAMILY: Asclepiadaceae

HABIT: Vine, to 15 ft.

HABITAT: Ornamental.

NATIVE TO: Indian Ocean, Malaya to Madagascar.

LEAF: Simple, opposite, 2 to 4 in. long; stalk shorter than blade; blade elliptical, dark green, leathery, typically blunt at both ends but often with a distinct small point at the tip; the midrib is especially distinct.

FLOWER: Clustered, bright white, conspicuous, very fragrant; each flower elongate, 5-parted, waxy in texture, tubular, about half as long as the leaves, flaring into 5 petal lobes.

FRUIT: Elongate heavy, fleshy pods, to 5 in. long; green then brown, divided into two halves by a shallow longitudinal furrow.

STEM: Slow-growing, climbing vine, twining about itself and whatever else it encounters.

•Stephanotis is especially prized for bridal bouquets, corsages, and leis. Although the vine flowers over an extended time, it grows slowly under the best of circumstances and usually requires careful attention for success. For these reasons stephanotis, while almost a household word, is itself not commonly encountered in casual locations.

Strelitzia nicholai

Bird-of-paradise tree
White bird of paradise

FAMILY: Strelitziaceae

HABIT: Small tree, to about 20 ft.

HABITAT: Ornamental.

NATIVE TO: South Africa.

LEAF: Similar to the next, but much larger, coarser; leaf long-stalked, to 8 ft. or more, blade about half as long as stalk; borne on a trunk rather than from the ground; the leaf cluster is two-ranked, making a fan.

FLOWER: The inflorescence of 2 to 3 large flowers is similar to the next, except short-stalked (buried in the fan of leaves) and the bract is blue-black or purple, the sepals pure white or reddish, and the arrow-shaped tongue pale blue.

STEM: The trunk is slender and usually branched near the ground, so the trunks typically appear in clumps. They are not woody like Ravenala, but like that of banana (Musa).

•The plant looks like a small traveler's palm (Ravenala), which see, in its fan-like leaf arrangement. The leaves are banana-like. In windy locations they become torn along the lateral veins (illustrated lower right) as in banana. All three of these genera are closely related.

Strelitzia alba is similar but somewhat smaller, and differs in some details of the flower structure.

Strelitzia reginae

Bird of paradise
Strelitzia

FAMILY: Strelitziaceae

HABIT: Perennial herb, to 6 ft.

HABITAT: Ornamental.

NATIVE TO: South Africa.

LEAF: Clustered from the ground; conspicuously and regularly pinnate-veined; the midvein may be red. The blade is paddle-shaped, 6 in. wide by 1½ ft. long; with edges somewhat rolled upward, and a bluish cast; the stalk is longer than the blade; the whole leaf to 5 ft. tall. The leaves may be inconspicuously two-ranked right at the ground.

FLOWER: About 6 flower buds are encased in a horizontal boat-shaped green bract (modified leaf), to 8 in. long, often colored with blue or red; opening upwards from the bract one by one over several days. Each has elongate orange petals and a complex elongate blue "tongue" at the center, composed of petals, stamens, and style.

FRUIT: Small, 3-chambered pod.

•A striking and unmistakable flower with easily imagined ornithological affinities, this is sometimes grown as a pot plant. However, it is slow-growing, requiring as much as seven years to reach flowering. It is most commonly seen in elegant (expensive) floral arrangements.

Strongylodon macrobotrys

Jade vine

FAMILY: Leguminosae, Faboideae

HABIT: Rampant hanging, climbing vine to 60 ft.

HABITAT: Ornamental.

NATIVE TO: Philippines.

LEAF: Alternate, 1x odd-pinnate with only 3 leaflets close together at the end of a long stalk, each elliptical, glossy, pointed, to 5 in. long by half as wide; opening light green or reddish, turning dark green.

FLOWER: Hanging inflorescence, each to 4 ft. long, with up to 100 large, exaggerated pea-like flowers, each 2 to 3 in. long; petals soft, velvety, curved back and up, an unusual bluish jade-green; arising from a small purplish calyx tube; on short purple stems.

FRUIT: Swollen, non-opening pod, containing 3 to 10 seeds.

STEM: Woody, attaining 1 in. in diameter.

•The inflorescences are large and hanging, but despite their size, not conspicuous at a distance because of their greenish color. The individual flowers are borne horizontally, with a curved, upturned aspect. The inflorescence of this plant is striking when viewed close-to. Its color is nearly unique among plants, the flower looks artificial, and as though defying gravity with its strongly upturned tip.

Suriana maritima

Bay cedar

FAMILY: Simaroubiaceae

HABIT: Shrub, to 10 ft.

HABITAT: Seashores.

NATIVE TO: Tropical shores worldwide, possibly spreading from the New World.

LEAF: Simple, alternate but often clustered, small, narrow, somewhat thick, with parallel edges, to 1 1/2 in. long by 1/4 in. wide; densely hairy.

FLOWER: Single, or few-clustered; small, yellow, with 5 prominent green sepals opening to display 5 spreading blunt, equal petals surrounding a cluster of stamens and style, the whole less than 1/2 in. across.

FRUIT: Tiny hairy pod, with 5 longitudinal angles.

STEM: Red-brown, the younger densely hairy.

•When growing in locations on lee shores sheltered from wind, bay cedar can attain the size of a small tree. Its wood is extremely dense, close grained, and reddish in color, useful in native cultures for fashioning fish hooks and spears. Despite its common name, bay cedar, an angiosperm, is not related to other cedars which belong to the gymnosperms.

Swietenia mahogoni

Mahogany
West Indies mahogany.

FAMILY: Meliaceae

HABIT: Majestic tree, to 75 ft.

HABITAT: Rain forest; shade tree.

NATIVE TO: West Indies.

LEAF: Alternate, 1x even-pinnate, 3 to 10 pairs of shiny, stiffish, glossy leaflets, each about 6 in. long; blades sharp-pointed; short-stalked; midvein distinct and gently curved.

FLOWER: Small greenish-yellow flowers in elongate clusters, 6 in. long; fragrant. Summer flowering.

FRUIT: Erect, hard, wooden, green then brown, pear-shaped, to to 5 in.long (first illustration); which splits into 5 segments containing winged seeds hanging from a central stalk, about the size and shape of maple keys (second illustration).

STEM: The trunk may have low buttresses at the base. The bark is longitudinally ridged and somewhat scaly. The trunk of an ancient mahogany is illustrated lower right. Most trees of this size have already been cut for lumber.

•The dark red wood of this tree was prized as the original mahogany of commerce, and is still used for furniture, etc. Most mahogany lumber now comes from Swietenia macrophylla of South America. West Indian mahogany trees get a little "bald" during the dry season, but new leaves come before the old ones are all gone. This species is a very popular shade tree. Where protected against lumbering, trees eventually assume a majestic size and appearance.

Northerners familiar with maple keys will be surprised to find a duplicate in the tropics. Although maples and mahoganies are unrelated, both have evolved strikingly similar winged seeds as an effective method of dispersing seedlings from under the parent tree.

Syngonium podophyllum (syn. Nephthytis spp.)

Nephthytis
Arrowhead vine

FAMILY: Araceae

HABIT: Woody, climbing vine or sprawling shrub.

HABITAT: Ornamental.

NATIVE TO: Central America.

LEAF: Simple, alternate, stalk clasping; to 2 ft. long. Blade unusual; divided into 5 to 9 parts grading smaller from the central one to the laterals (see illustration).

FLOWER: Typical spathe and spadix of the family (see Anthurium). Spathe in two parts: green, swollen, and unopening at base, and a whitish hood above, the whole about 3 to 4 in. long. Spadix thick-columnar, composed of numerous small, radiating and regularly ordered flowers.

STEM: Viney, woody, gray-barked, commonly bearing roots (which are used in climbing) at the nodes.

•Seedlings and small plants bear simple arrow-shaped leaves. As the plant ages, the leaf form changes. The lower lobes become larger and completely separate from the upper part of the leaf. They in turn become lobed on one side, and the process repeats until the leaf blade consists of 5 to 9 mostly fully separated units. In some areas, Syngonium is used as a ground cover, kept down by occasional pruning (illustration top right).

Tabebuia rosea (syns. T. pentaphylla, Tecoma pentaphylla)

Pink poui
Trumpet tree
West Indies cedar
Pink tecoma

FAMILY: Bignoniaceae

HABIT: Tree, to 60 ft.

HABITAT: Ornamental.

NATIVE TO: Central and South America.

LEAF: Opposite, palmately compound, typically with 3 to 5 leaflets; leaflets elliptical, glossy, short-stalked; to 6 in. long; deciduous.

FLOWER: Single or loosely clustered, 5-parted; a slightly upturned white tube to 3 1/2 in. long; flaring petal lobes crepe-like, faded purple-pink, to white.

FRUIT: Cylindrical pod, to 1 1/2 ft., green (see flower illustration), turning brown, containing papery seeds.

STEM: Trunk may be somewhat fluted or buttressed at the bottom. The smaller branches tend to have a graceful curving pattern.

•This tree has erratic blooming. The main show of flowering occurs when the tree is almost bare of leaves. Then it has a significant period of blooming, when it is flushed with color. At other times a few blossoms are usually scattered among its branches at any season. It provides excellent building timber, like oak, which resists termites. It can grow to become one of the tallest trees in the Caribbean, and is a common shade tree.

There are many additional common species of Tabebuia in the tropics. Tabebuia pallida and T. serratifolia are described on the next page.

Tabebuia serratifolia
Tabebuia pallida

Yellow poui
White poui
White cedar

FAMILY: Bignoniaceae

•Illustrated opposite are two more of the several common species of Tabebuia (see previous page).

Tabebuia serratifolia, yellow poui, illustrated upper left, has yellow flowers as the common name implies. The palmately compound leaves usually have five leaflets with distinctly toothed edges. This species is native to northern South America to the West Indies.

Tabebuia pallida, white poui or white cedar, (the next three illustrations), has delicate, scattered individual white or pale lilac flowers with a slightly yellow throat. The leaflets are usually three, and the species is native to the West Indies. One of the illustrations shows the flower; the other two show the seed pod and winged seeds.

Another common species is Tabebuia argentia (not illustrated), silver poui, which has yellow flowers. The foliage is distinctly silvery in color because of silvery scales on the scurfy leaf blades.

Tabernaemontana divaricata (syn. Ervatamia coronaria)

Cape jasmine
Cape gardenia
Butterfly gardenia
Chinese gardenia

FAMILY: Apocynaceae

HABIT: Shrub, to 8 ft.

HABITAT: Ornamental.

NATIVE TO: India.

LEAF: Simple, in opposite pairs, one usually distinctly larger than the other; 3 to 6 in. long; stiff, often shiny.

FLOWER: Single or clustered, white, tubular, 5-parted, to 1½ in. across; sepals small, inconspicuous; petals opening into 5 separate, narrow, elongate lobes (single form) or into more numerous, broader overlapping lobes (double variety); in both cases petals are crimped, waxy, white, fragrant at night. Single and double varieties are illustrated.

FRUIT: Pod with a curved beak, 1 to 3 in. long, in pairs; scarlet inside.

STEM: Much branched.

•This white-flowered shrub can be confused with gardenias, jessamines, jasmines, etc. In this one, look especially for the five narrow petal lobes in the single-flowered form. These lobes are separate almost to the base of the petal and slightly spiral in placement. The petals are distinctly crimped or wrinkled at the edges in the single form, or almost throughout in some doubled varieties. The unequal pairing of leaves also helps in identification, though it is not absolutely dependable.

Tamarindus indica

Tamarind

FAMILY: Leguminosae, Caesalpinioideae

HABIT: Majestic light green tree, to 80 ft.

HABITAT: Ornamental; naturalized.

NATIVE TO: Asia, Africa.

LEAF: Alternate, 1x even-pinnate; 10 to 18 pairs of very regular leaflets, each to 1 in. long, with parallel edges.

FLOWER: Few-flowered clusters; dark red buds open into pale yellow legume flowers, about 1 in. across. Flowers in summer.

FRUIT: Lumpy round pod, an inch or more in diameter, by 2 to 6 in. long; irregularly constricted, usually curved, cinnamon brown.

STEM: The bark is rough.

•The delicate compound foliage gives these large trees a light green feathery aspect. Tamarind was brought to the Caribbean in the 1600's. It is now widely dispersed as a shade tree and grown also for its fruit and tight-grained, durable wood. The latter is an excellent cabinet material. The acid pulp of the fruit can be eaten raw, but more commonly is made into a beverage or used for chutneys, curries, candy, and preserves.

Tecoma stans (syn. Stenolobium stans)

Ginger thomas
Yellow elder
Yellow trumpet tree

FAMILY: Bignoniaceae

HABIT: Shrub or small tree, to 20 ft.

HABITAT: Ornamental; naturalized, roadsides particularly.

NATIVE TO: West Indies and tropical Americas.

LEAF: Alternate, 1x odd-pinnate, to 10 in. long; leaflets 5 to 13, blades broadest below the middle, narrowing evenly to a sharply pointed tip, to 4 in. long, with coarsely toothed edges.

FLOWER: Few-flowered clusters; flowers tubular, bright yellow, 5-parted; corolla tube narrow, then flaring sharply a half inch or so above the calyx; petal lobes wavy, 3 down, 2 up.

FRUIT: Narrow, hanging, cylindrical pod, light green, then brown, to 8 in.

•The tree flowers mostly in fall and fruits in winter. The pods are often produced copiously and are conspicuous, like hanging fingers, in the trees. They have two compartments lengthwise and contain large flat seeds with papery wings. Tecoma stans is a prolific bloomer as well as a rapid grower. It volunteers along roadsides and waste areas as well as in yards and fields all of which it graces with its striking sunny yellow. It has been designated the official flower of the U. S. Virgin Islands and the Bahamas.

Tecomaria capensis

Cape honeysuckle

FAMILY: Bignoniaceae

HABIT: Shrub, to 6 ft., sometimes sprawling.

HABITAT: Ornamental.

NATIVE TO: South Africa.

LEAF: Opposite, 1x odd-pinnate, to 6 in. long, with 5 to 9 toothed or incut, glossy, broadly elliptical or sometimes almost round leaflets.

FLOWER: Clustered; bright red or orange-red, narrowly tubular; petal tube 2 in. long, slightly bent, opening into 5 lobes, with 4 arching stamens and style protruding.

FRUIT: An elongate pod.

•When Tecomaria is growing in a good location, the contrast between the dark green foliage and the conspicuous fiery red inflorescences of this plant is striking. It is a favorite ornamental for this reason and also because of its easy propagation and rapid growth. There is also a yellow-colored variety. Both can be used as hedges.

The flowers of Tecomaria capensis can be confused with those of Pyrostegia ignea, flame vine, but the latter has compound leaves with only two or three leaflets and is more distinctly a vine, climbing by tendrils. Tecomaria lacks tendrils.

Terminalia cattapa

Indian almond
Seaside almond
Tropical almond
Almond

FAMILY: Combretaceae

HABIT: Tree, to 80 ft.

HABITAT: Naturalized and planted as a seaside shade tree.

NATIVE TO: Centered in the Malay Peninsula.

LEAF: Simple, alternate, large, stiff, crowded at the branch tips, broader outward, blunt-tipped (pointed in saplings), 5 to 6 in. across, 10 in. long; pinnately veined; leaf stalk very short.

FLOWER: Small, green-yellow, inconspicuous, in spikes at branch tips (illustrated).

FRUIT: Ovoid, slightly 2-ribbed, hard, green, then yellow or reddish, 2 in. long, containing one central wooden nut in a fleshy fibrous covering. The seed within is oily and edible, flavored like an almond.

STEM: Often a conspicuous horizontal branching pattern, especially when young.

•The leaves usually turn red before they fall. This northern characteristic is unusual among tropical trees and is a good aid to recognition, In most places, this happens leaf by leaf, scattered fashion among the foliage (illustration upper middle right), but sometimes a tree is found in which most of the leaves have turned red nearly at once (illustration lower right). Terminalia withstands salt spray well and is widely planted near seashores.

Theobroma cacao

Cocoa
Cacao
Chocolate tree

FAMILY: Byttneriaceae

HABIT: Small tree, to 25 ft.

HABITAT: Cultivated.

NATIVE TO: Central and tropical South America.

LEAF: Simple, alternate, thin-leathery, shiny, very broadly elliptical, to 12 in. long; short-stalked, often distinctly hanging; surface crinkly, reddish when young; not deciduous.

FLOWER: Tiny, long stemmed, borne directly from the trunk or main branches, white or pink and yellow, less than 1/2 in. across, inconspicuous.

FRUIT: Hanging 10-ribbed, greenish, red, or dark red swollen pod, to 10 in. long by 3 or 4 in. thick; tapering ends; seeds numerous, ellipsoid, in mucilaginous pulp, to 1 in. in size.

•This tree was cultivated by the Aztecs. Columbus took seeds back to Europe in 1502. It is now cultivated extensively in Tobago and Grenada especially. Cocoa requires some shade, usually provided by a companion planting. Trees bear at four years and flower several times a year. The pods are opened and the seeds removed. Pulp is sweated off the seeds by fermentation. The seeds are then dried, roasted, de-shelled, and ground into an oily paste (raw bitter chocolate). This contains a stimulant related to caffeine. Cacao is the tree, cocoa the extract of the seeds. Coca (the source of cocaine) is an entirely different shrub, Erythroxylum coca.

Thespesia populnia

Oti haiti
Portia
Milo

FAMILY: Malvaceae

HABIT: Tree, to 50 ft.

HABITAT: Seaside strands.

NATIVE TO: West Indies.

LEAF: Simple, alternate, heart-shaped, glossy, pointed; long-stalked; 3 to 7 in. long.

FLOWER: Single, in leaf axils, to 2 in. across; a shallow bell of 5 pale yellow petals, reddish in bud and at base, crepe-like, imbricated, short-lived, Hibiscus-like; fading to deep dull pink or orange (illustrated lower middle right). The central column of stamens and style is like that of Hibiscus, but squat and thick, not protruding.

FRUIT: Erect flattened sphere, to 2 in. wide, dry, somewhat dimpled, marked with 5 indistinct longitudinal ridges; containing a few large brown seeds The fruits are dusty black when mature.

STEM: The tree has especially spreading lower branches when growing in the open.

•The wood is useful for food dishes and containers because it is easily worked and imparts no taste or odor of its own. Because of its resistance to salt spray, Thespesia is sometimes planted as a shade tree on seaside properties. See also the similar Hibiscus tiliaceus.

Thevetia peruviana (syn. T. neriifolia)

Yellow oleander
Be-still tree

FAMILY: Apocynaceae

HABIT: Shrub or small tree, to 20 ft.

HABITAT: Ornamental; sometimes a clipped hedge.

NATIVE TO: Tropical America.

LEAF: Simple, alternate, very narrowly elliptical, glossy above, light green, about 1/4 in. wide by up to 6 in. long; no separate leaf stalk; not deciduous.

FLOWER: Single or few-clustered, tubular, yellow or orange-yellow, 5-parted, trumpet-like, spreading, to 2 in. across, fragrant.

FRUIT: Squat-spherical, slightly 4-angled like a bell pepper, green, shrinking and darkening to almost black when ripe (illustrated); containing 1 or 2 large, flattened, smooth brown seeds ("lucky nut").

•The plant contains a milky sap and all parts are highly poisonous (whence the second common name.) See also oleander (Nerium oleander) of the same plant family. These two can be separated readily by flower color and shape of fruit. Thevetia can also be separated easily from yellow allamanda (Allamanda cathartica) which also has tubular yellow flowers, by the broad leaves in the latter.

Thunbergia grandiflora

Sky flower
Blue trumpet vine
Purple allamanda
Thunbergia

FAMILY: Acanthaceae

HABIT: Twining woody vine, or partly shrubby, to 50 ft.

HABITAT: Ornamental.

NATIVE TO: India.

LEAF: Simple, opposite, short-stalked, to 8 in. long; blade rough, 3 to 5-veined from base, angular, coarse-toothed, sometimes lobed and spearlike, broader at the base (often heart-shaped), narrowing to a point.

FLOWER: Solitary or few-clustered; velvety, tubular, blue or purplish, with a pale creamy throat, 5-parted, about 3 in. long and wide; bottom petal lobe slightly the largest.

•Each flower arises above a pair of bracts (small modified leaves). The calyx itself is small and ringlike, thus unusual and distinctive. The plant flowers nearly continuously. There is also a white flowered variety of Thunbergia grandiflora.

Several other species of Thunbergia may be encountered in gardens. T. laurifolia, laurel-leafed thunbergia, is one in which the flower color is sky blue and the leaves leathery, elliptical, and smooth-edged.

Thunbergia grandiflora can be confused in flower or name with members of the genera Cryptostegia and Allamanda, which see.

Thunbergia alata

Black-eyed Susan vine
Clock vine

FAMILY: Acanthaceae

HABIT: Perennial viney herb, climbing to 8 ft.

HABITAT: Ornamental; naturalized.

NATIVE TO: Tropical Africa.

LEAF: Simple, opposite, stalked, to 3 in. long; leaf-stalk more or less winged; blade heart-shaped (basal lobes round) or even spear-shaped (basal lobes pointed), usually coarsely toothed and rough surfaced; the 3 to 5 main veins originate at the point of stalk attachment.

FLOWER: Mostly single, long-stalked, 5-parted; bright orange-yellow with a sharply defined dark (almost black) center; the 5 almost equal petals flaring widely into a flat, shallow flower about 3 in. across, arising from a cup of green, sepal-like bracts. There are also white or cream colored varieties with or without a dark center.

STEM: Twining and viney, but lacking tendrils.

•The flowers give this persistent, rapidly-growing, almost weedy vine a delightfully cheerful aspect. The flowers, though not large, stand out like egg yolks against the dark background of the foliage. The petals are somewhat cupped as they open, then flatten back, and finally continue to bend backward behind the plane of the flower as it ages.

Tibouchina urvilleana

Glory bush
Lasiandra
Purple glory tree

FAMILY: Melastomataceae

HABIT: Shrub, to 15 ft.

HABITAT: Ornamental, naturalized.

NATIVE TO: Brazil.

LEAF: Simple, opposite, short-stalked; blade broadly elliptical, pointed, to 6 in. long; with 3 to 5 main veins running lengthwise from base, secondary veins crosswise between. Old leaves may turn bright red before dropping.

FLOWER: Single or few-clustered, very conspicuous; each bright purple, velvety, almost flat, with 5 spreading equal petals, about 4 in. across, with 10 peculiarly angled stamens at the center; covered in bud by large reddish hairy sepals.

STEM: Four-sided, reddish when young.

•This is a striking shrub. The foliage has a silvery sheen from the finely hairy upper surfaces of the leaves, and the large flowers are almost painfully intense in color in the sun. Other related species are similar, especially in the somewhat unusual leaf, but have white or pink flowers.

Tillandsia usneoides

Spanish moss
Graybeard

FAMILY: Bromeliaceae

HABIT: Clumps or hanging festoons to 20 ft.

HABITAT: Grows entirely in the air, hanging on trees, telephone poles, wires, etc.

NATIVE TO: Tropical America.

LEAF: Scattered, narrow, stalkless, clasping, gray-green, to 2 in. long; usually curved or bent.

FLOWER: Small, greenish or bluish, inconspicuous among the leaves.

STEM: Extensive, slender ($^1/_{16}$ in.), thread-like, elongate, greenish, eventually tangled, twisted, and twining. The radiating structures in the illustration at center right are young stems with large buds.

•All parts of the plant are covered with silvery gray scales. This is a protection against light and dessication. Protection against drying is particularly important because Spanish moss has no roots. It takes its water from the air as it can. The gray-green color of the vegetation thus is not an indication of parasitism. Spanish moss is not parasitic. Technically an epiphyte (see description at Ficus spp.) it uses its host only as a place to live.

The plant was formerly harvested for use as stuffing. It is also useful for fodder.

Veitchia merrillii

Christmas palm
Manila palm

FAMILY: Palmae

HABIT: Small tree, to 20 ft.

HABITAT: Ornamental.

NATIVE TO: Philippines.

LEAF: Alternate, 1x pinnate, to 6 ft. long; ascending and arching, older hanging; leaflets numerous (50 to 60 each side), narrowing at both ends.

FLOWER: Inflorescences clustered on spikey stems at the junction between the gray trunk and green clasping leaf base; branched and rebranched; sexes separated. Male flowers, with numerous radiating stamens, outward (illustrated); female flowers inward, near trunk.

FRUIT: Numerous, conspicuous, initially white, becoming bright red; egg-shaped, each about 1 in. long.

•This is one of many common species of small feather palms (see comments at Latania). This one is especially prized as a compact ornamental because of the bright red fruits ripening near Christmas time. Its relatively small size recommends it for lawns and yards where it is frequently planted.

Wedelia trilobata

Wedelia

FAMILY: Compositae

HABIT: Herbaceous perennial.

HABITAT: Cultivated ground cover.

NATIVE TO: Tropical America.

LEAF: Simple, opposite, variable, but short-stalked when fully expressed, 2 to 5 in. long; blade broadly triangular, glossy; edges may be toothed and incut with two large teeth or pointed lobes at the base.

FLOWER: Solitary; small daisy-like heads, yellow throughout or orange-yellow at the center yellowing outward, about 1 in. across; everblooming.

STEM: Trailing to 3 or 4 ft., somewhat fleshy, rooting at the nodes, forming a solid cover 1 to 2 ft. high.

•This is one of few members of the composite family (Compositae) to be found in the tropics as compared with northern latitudes where composites are common. Wedelia is widely planted as a ground cover or lawn that does not have to be mowed. It tolerates full sun or partial shade, and can be clipped to enhance uniformity (as shown in the top left illustration). See also Senecio confusus, another tropical composite with which Wedelia might be confused.

Yucca aloifolia

Spanish bayonet
Yucca

FAMILY: Liliaceae

HABIT: Shrub or tree, to 25 ft.

HABITAT: Ornamental, naturalized.

NATIVE TO: Tropical Americas.

LEAF: Linear, stiffish, not fleshy, very sharp-pointed and hard surfaced (see illustration); margins very finely (almost invisibly) toothed but not thread-bearing; about 2½ ft. long by 2½ in. wide; the living leaves usually clustered at the tip of the slowly growing branches.

FLOWER: Erect, terminal, open, densely-flowered clusters; to 2 ft. high; flowers nodding, white, tubular, 6-parted, lily-like; each to 4 in. across.

STEM: Slow growing, single or branched, making a large trunk with time.

•The leaves that arise along the stem die but are somewhat persistent if not removed. The stem increases in size slowly, eventually becoming a trunk a foot or more in diameter. There are several named horticultural varieties of this species, as well as many other species of Yucca. Some Yuccas lack a trunk. Some have fibers, like strings, hanging from the leaf edges. Do not confuse yucca with the foodstuff called yuca (which see). Yucca flowers can be eaten as a crisp and colorful addition to salads.

Yucca aloifolia is often planted to form an inhospitable hedge at property boundaries, and can also be used to fence pastures effectively. Yucca barriers are kept low by cutting out the terminal bud periodically. This induces suckering lower down.

INDEX TO COMMON NAMES

For each common name, a scientific name is given to the right. These scientific names are the alphabetical page headings in the text. To look one up, you usually need remember only the first four letters of the scientific name.

Common name	Scientific name
Black pepper	Piper nigrum
Black-eyed Susan	Thunbergia alata
Bleeding glory bower	Clerodendrum thomsoniae
Bleeding heart	Clerodendrum thomsoniae
Blimbing	Averrhoa bilimbi
Blue palm	Latania loddigesii
Blue trumpet vine	Thunbergia grandiflora
Bo tree	Ficus religiosa
Boat lily	Rhoeo spathacea
Bottlebrush	Callistemon speciosus
Bougainvillea	Bougainvillea glabra
Bougainvillea	Bougainvillea spectabilis
Bowstring hemp	Sansevieria trifasciata
Brazilian pepper tree	Schinus terebenthifolius
Breadfruit	Artocarpus altilis
Breadnut	Artocarpus altilis
Bridal bouquet	Stephanotis floribunda
Buddha's lamp	Mussaenda spp.
Buddhist pine	Podocarpus macrophyllus
Bull hoof	Bauhinia purpurea
Bull hoof	Bauhinia variegata
Bullock's heart	Annona reticulata
Bush morning glory	Ipomoea fistulosa
Buttercup tree	Cochlospermum vitifolium
Butterfly gardenia	Tabernaemontana divaricata
Butterfly pea	Clitoria ternatea
Buttonwood	Conocarpus erecta
C	
Cacao	Theobroma cacao
Cactus	Cephalocereus royenii
Cajeput	Melaleuca quinquenervia
Calabash tree	Crescentia cujete
Calamondin orange	Citrifortunella x mitis
Candelabra cactus	Euphorbia lactea
Candlebush	Cassia alata
Candlebush	Cassia didymobotrya
Candlestick bush	Cassia alata
Cannonball tree	Couroupita guianensis
Cape honeysuckle	Tecomaria capensis
Cape leadwort	Plumbago auriculata
Carambola	Averrhoa carambola
Caricature plant	Graptophyllum pictum
Carissa	Carissa grandiflora
Carrion flower	Stapelia gigantea
Casha	Acacia tortuosa
Cassava	Manihot esculenta
Castor bean	Ricinus communis
Casuarina	Casuarina equisetifolia
Century plant	Agave americana
Century plant	Agave missionum
Ceriman	Monstera deliciosa
Chalice vine	Solandra maxima
Chenille plant	Acalypha hispida
Cherimoya	Annona cherimola
Chili pepper	Capsicum sp.
Chinese gardenia	Tabernaemontana divaricata
Chinese hat	Holmskoldia sanguinea
Chinese hibiscus	Hibiscus rosa-sinensis
Chocolate tree	Theobroma cacao
Christmas bush	Euphorbia leucocephala

Christmas candle .... Cassia alata
Christmas candle .... Pedilanthus tithymaloides
Christmas palm .... Veitchia merrillii
Christmas star .... Euphorbia pulcherrima
Christmasberry tree .... Schinus terebenthifolius
Cinnamon .... Cinnamonium zeylanicum
Citrus fruits .... Citrus spp.
Clam-and-cherry .... Cordia dentata
Clockvine .... Thunbergia alata
Clustered fishtail palm .... Caryota mitis
Coca .... Erythroxylum coca
Cockspur coral tree .... Erythrina crista-gallii
Cocoa .... Theobroma cacao
Cocoa shade .... Gliricidia sepium
Coconut .... Cocos nucifera
Coconut palm .... Cocos nucifera
Coffee .... Coffea arabica
Common bamboo .... Bambusa vulgaris
Common jasmine .... Jasminum officinale
Cook's pine .... Araucaria columnaris
Copper leaf .... Acalypha wilkesiana
Coral blow .... Russelia equisetiformis
Coral hibiscus .... Hibiscus schizopetalus
Coral plant .... Jatropha multifida
Coral shower tree .... Cassia grandis
Coral tree .... Erythrina crista-gallii
Coral tree .... Erythrina variegata
Coral vine .... Antigonon leptopus
Coralita .... Antigonon leptopus
Cotton .... Gossypium hirsutum
Crab claw .... Erythrina variegata
Crab's eyes .... Abrus precatorius
Crape gardenia .... Tabernaemontana divaricata
Crape jasmine .... Tabernaemontana divaricata
Crossandra .... Crossandra infundibuliformis
Croton .... Codiaeum variegatum
Crown flower .... Calotropis procera
Crown-of-thorns .... Euphorbia milii
Cuban royal palm .... Roystonea regia
Cucumber tree .... Averrhoa bilimbi
Cup of gold .... Solandra maxima
Cup-and-saucer .... Holmskoldia sanguinea
Custard apple .... Annona reticulata
Custard apple .... Annona squamosa
Cycads .... Cycas spp.
Cypress vine .... Ipomoea quamoclit

D

Danish flag .... Clerodendrum thomsoniae
Dasheen .... Alocasia macrorrhiza
Dasheen .... Colocasia esculenta
Dead rat tree .... Adansonia digitata
Devil's gut .... Cuscuta spp.
Divi divi .... Caesalpinia coriaria
Dodder .... Cuscuta spp.
Dracena .... Cordyline terminalis
Dracena .... Dracaena deremensis
Dragon bones .... Euphorbia lactea
Dumbcane .... Dieffenbachia spp.
Dwarf lilyturf .... Ophiopogon japonicus
Dwarf poinciana .... Caesalpinia pulcherrima

E

Elephant's ear . . . . Alocasia macrorrhiza
Eranthemum . . . . Pseuderanthemum reticulatum

F

False bird of paradise . . . . Heliconia spp.
False hop . . . . Justicia brandegeana
False ironwood . . . . Casuarina equisetifolia
False mistletoe . . . . Phoradendron spp.
Fan palm . . . . Latania loddigesii
Fern palms . . . . Cycas spp.
Fiddleleaf fig . . . . Ficus lyrata
Fig . . . . Ficus elastica
Firecracker . . . . Russelia equisetiformis
Fishtail palm . . . . Caryota mitis
Flamboyant . . . . Delonix regia
Flame of the forest . . . . Spathodea campanulata
Flame tree . . . . Delonix regia
Flame vine . . . . Pyrostegia ignea
Flame-of-the-woods . . . . Ixora coccinea
Flamingo flower . . . . Anthurium andraeanum
Florida holly . . . . Schinus terebenthifolius
Flower fence . . . . Caesalpinia pulcherrima
Fountain bush . . . . Russelia equisetiformis
Frangipani . . . . Plumeria alba
Frangipani . . . . Plumeria obtusa
Frangipani . . . . Plumeria rubra
Fringed hibiscus . . . . Hibiscus schizopetalus

G

Galphimia vine . . . . Tristellateia australasiae
Gardenia . . . . Gardenia taitensis
Geiger tree . . . . Cordia sebestena
Geranium tree . . . . Cordia sebestena
Geranium-leafed aralia . . . . Polyscias guilfoylei
Giant crown flower . . . . Calotropis gigantea
Giant Indian milkweed . . . . Calotropis procera
Ginger lily . . . . Alpinia purpurata
Ginger thomas . . . . Tecoma stans
Glory bush . . . . Tibouchina urvilleana
Golden coconut . . . . Cocos nucifera
Golden eranthemum . . . . Pseuderanthemum reticulatum
Golden rain . . . . Cassia fistula
Golden shower . . . . Cassia fistula
Golden trumpet . . . . Allamanda cathartica
Good luck plant . . . . Cordyline terminalis
Gourd tree . . . . Crescentia cujete
Gout plant . . . . Jatropha podagrica
Granadina . . . . Passiflora spp.
Grapefruit . . . . Citrus paradisi
Graybeard . . . . Tillandsia usneoides
Green ebony . . . . Jacaranda mimosifolia
Guava . . . . Psidium guajava
Guinea tamarind . . . . Adansonia digitata
Gumbo limbo . . . . Bursera simaruba

H

Hala . . . . Pandanus odoratissimus
Hala . . . . Pandanus utilis
Hanging heliconia . . . . Heliconia spp.
Hat rack cactus . . . . Euphorbia lactea
Hawaiian hibiscus . . . . Hibiscus rosa-sinensis

Heliconia.................................Heliconia spp.
Hell weed ....Cuscuta spp.
Hibiscus ....Hibiscus rosa-sinensis
Horseradish tree ....Moringa pterygosperma
Hunter's robe ....Epipremnum aureum
Hura ....Hura crepitans
Hydrangea tree...........................Dombeya wallichii

I

India rubber vine........................Cryptostegia grandiflora
Indian almond............................Terminalia cattapa
Indian banyan ....Ficus benghalensis
Indian coral tree ....Erythrina variegata
Indian laburnum ....Cassia fistula
Indian mulberry ....Morinda citrifolia
Indian rubber tree.......................Ficus elastica
Indian tree..............................Euphorbia tirucalli
Ironwood.................................Casuarina equisetifolia
Ixora ....Ixora coccinea

J

Jacaranda................................Jacaranda mimosifolia
Jack fruit ....Artocarpus heterophyllus
Jackwood ....Cordia dentata
Jacob's coat ....Breynia disticha
Jade vine ....Strongylodon macrobotrys
Japanese hibiscus ....Hibiscus schizopetalus
Japanese lantern ....Hibiscus schizopetalus
Japanese poinsettia......................Pedilanthus tithymaloides
Japanese yew ....Podocarpus macrophyllus
Jasmine ....Jasminum officinale
Jerusalem thorn..........................Parkinsonia aculeata
Jessamine ....Jasminum officinale
Joseph's coat ....Acalypha wilkesiana
Jungle geranium ....Ixora coccinea

K

Kalamona ....Cassia surattensis
Kalanchoe................................Kalanchoe blossfeldiana
Kapok ....Ceiba pentandra
Kino ....Coccoloba uvifera
Kohio vine...............................Ipomoea horsfalliae
Kumquat ....Fortunella spp.

L

Lantana..................................Lantana camara
Lasiandra ....Tibouchina urvilleana
Latan palm ....Latania loddigesii
Laurel-leaved thunbergia ....Thunbergia laurifolia
Leadwort.................................Plumbago auriculata
Lemon....................................Citrus limon
Life plant ....Kalanchoe blossfeldiana
Lignum vitae ....Guaiacum officinale
Lime ....Citrus aurantifolium
Lipstick tree ....Bixa orellana
Livi divi ....Caesalpinia coriaria
Lobster claw ....Heliconia spp.
Love vine ....Cuscuta spp.

M

Mace ....Myristica fragrans
Madagascar jasmine ....Stephanotis floribunda
Madagascar periwinkle ....Catharanthus roseus

Common name	Scientific name
Madre de cacao	Gliricidia sepium
Mahoe	Hibiscus tiliaceus
Mahogany	Swietenia mahogoni
Maltese cross	Ixora coccinea
Mammee apple	Mammea americana
Manchineel	Hippomane mancinella
Mandarin orange	Citrus reticulata
Mango	Mangifera indica
Manila palm	Veitchia merrillii
Manioc	Manihot esculenta
Match-me-if-you-can	Acalypha wilkesiana
Maypole	Agave americana
Maypole	Agave missionum
Medicinal aloe	Aloe barbadense
Medinella	Medinella magnifica
Melon tree	Carica papaya
Mesapple	Manilkara zapota
Mesple	Manilkara zapota
Mexican creeper	Antigonon leptopus
Mexican flame vine	Senecio confusus
Mexican love chain	Antigonon leptopus
Mexican shrimp plant	Justicia brandegeana
Mickey-mouse plant	Ochna serratula
Milk bush	Euphorbia tirucalli
Milo	Thespesia populnia
Mistletoe	Phoradendron spp.
Mondo grass	Ophiopogon japonicus
Monkey bread tree	Adansonia digitata
Monkey pistil	Hura crepitans
Monkey pod tree	Samanea saman
Monkey puzzle	Euphorbia lactea
Monkey puzzle tree	Araucaria araucana
Monkey's tail	Acalypha hispida
Monstera	Monstera deliciosa
Morning glory	Merremia tuberosa
Moses-in-a-cradle	Rhoeo spathacea
Mother-in-law plant	Dieffenbachia spp.
Mother-in-law's tongue	Albizia lebbeck
Mother-in-law's tongue	Sansevieria trifasciata
Mountain ebony	Bauhinia purpurea
Mountain ebony	Bauhinia variegata
Mussaenda	Mussaenda spp.

N

Common name	Scientific name
Naseberry	Manilkara zapota
Natal plum	Carissa grandiflora
Nephthytis	Syngonium podophyllum
Night-bloomnig cereus	Hylocereus undatus
Nodding hibiscus	Malaviscus arboreus
Norfolk Island pine	Araucaria heterophylla
Nutmeg	Myristica fragrans

O

Common name	Scientific name
Ochna	Ochna serratula
Octopus tree	Brassaia actinophylla
Oilcloth flower	Anthurium andraeanum
Old maid	Catharanthus roseus
Old world mistletoe	Viscum album
Oleander	Nerium oleander
Orange	Citrus aurantium
Orange glow vine	Senecio confusus

Common name	Scientific name
Orchid tree	Bauhinia purpurea
Orchid tree	Bauhinia variegata
Orchids	Orchids
Organ pipe cactus	Cephalocereus royenii
Oti haiti	Thespesia populnia
Oyster plant	Rhoeo spathacea

P

Common name	Scientific name
Pagoda flower	Clerodendrum paniculatum
Pain killer	Morinda citrifolia
Palma christi	Ricinus communis
Papaya	Carica papaya
Paperbark tree	Melaleuca quinquenervia
Paperflower	Bougainvillea glabra
Paperflower	Bougainvillea spectabilis
Parrot's plantain	Heliconia spp.
Pascuita	Euphorbia leucocephala
Passion flower	Passiflora spp.
Pawpaw	Carica papaya
Peepul tree	Ficus religiosa
Pencil tree	Euphorbia tirucalli
Pepper	Capsicum sp.
Pepper	Piper nigrum
Peppertree	Schinus terebenthifolius
Peregrina	Jatropha integerrima
Periwinkle	Catharanthus roseus
Peruvian pepper tree	Schinus molle
Physic nut	Jatropha multifida
Pine	Ananas comosus
Pineapple	Ananas comosus
Pink ball tree	Dombeya wallichii
Pink porcelain lily	Alpinia zerumbet
Pink poui	Tabebuia rosea
Pink shower tree	Cassia spp.
Pink tecoma	Tabebuia rosea
Pink trumpet vine	Podraena ricasoliana
Pink-and-white shower tree	Cassia spp.
Pipe organ cactus	Cephalocereus royenii
Pitch apple	Clusia rosea
Plantain	Musa x paradisiaca
Platterleaf	Coccoloba uvifera
Plumbago	Plumbago auriculata
Plumeria	Plumeria alba
Plumeria	Plumeria obtusa
Plumeria	Plumeria rubra
Podocarp	Podocarpus macrophyllus
Poet's jessamine	Jasminum officinale
Poi	Alocasia macrorrhiza
Poi	Colocasia esculenta
Poinciana	Delonix regia
Poinsettia	Euphorbia pulcherrima
Poison apple	Hippomane mancinella
Pomegranite	Punica granatum
Poor man's orchid	Bauhinia purpurea
Poor man's orchid	Bauhinia variegata
Portia	Thespesia populnia
Pothos	Epipremnum aureum
Pothos vine	Epipremnum aureum
Powder puff	Calliandra haematocephala
Precatory bean	Abrus precatorius
Prickly pear	Opuntia repens

Common name	Scientific name
Prickly pear	Opuntia rubescens
Pride of Barbados	Caesalpinia pulcherrima
Prince (or Prince's) vine	Ipomoea horsfalliae
Punk tree	Melaleuca quinquenervia
Purple allamanda	Cryptostegia grandiflora
Purple allamanda	Thunbergia grandiflora
Purple glory tree	Tibouchina urvilleana
Purple heart	Setcreasea pallida
Purple queen	Setcreasea pallida
Purple wreath	Petrea volubilis
Purple-leaved spiderwort	Rhoeo spathacea

Q

Common name	Scientific name
Queen's wreath	Petrea volubilis
Queen-of-the-night	Hylocereus undatus
Queensland umbrella tree	Brassaia actinophylla
Quick stick	Gliricidia sepium

R

Common name	Scientific name
Railroad vine	Ipomoea pes-caprae
Rain of gold	Galphimia glauca
Raintree	Samanea saman
Red flag	Mussaenda spp.
Red ginger	Alpinia purpurata
Red hot cattail	Acalypha hispida
Red leadwort	Plumbago indica
Red mangrove	Rhizophora mangle
Red powder puff	Calliandra haematocephala
Redbird	Pedilanthus tithymaloides
Ringworm bush	Cassia alata
Rosary pea	Abrus precatorius
Rose-of-Sharon	Hibiscus syriacus
Rosebay	Nerium oleander
Round calabash	Crescentia cujete
Royal palm	Roystonea regia
Royal poinciana	Delonix regia
Rubber	Hevea spp.
Rubber plant	Ficus elastica
Rubber vine	Cryptostegia grandiflora

S

Common name	Scientific name
Sacred tree	Ficus religiosa
Sago palm	Cycas revoluta
Saman	Samanea saman
Sandbox	Hura crepitans
Sandpaper vine	Petrea volubilis
Sansevieria	Sansevieria trifasciata
Sapodilla	Manilkara zapota
Sapota	Manilkara zapota
Sausage tree	Kigelia pinnata
Scaevola	Scaevola frutescens
Schefflera	Brassaia actinophylla
Scotch attorney	Clusia rosea
Scrambled eggs	Cassia surattensis
Screw pine	Pandanus odoratissimus
Screw pine	Pandanus utilis
Sea grape	Coccoloba uvifera
Sea purslane	Sesuvium portulacastrum
Seaside almond	Terminalia cattapa
Seaside morning glory	Ipomoea pes-caprae
Senna	Cassia spp.
Shaddock	Citrus grandis

Shakshak . . . . .Albizia lebbeck
She oak . . . . .Casuarina equisetifolia
Shell ginger . . . . .Alpinia zerumbet
Shower of gold . . . . .Cassia fistula
Shower of gold . . . . .Galphimia glauca
Shower tree . . . . .Cassia spp.
Shrimp plant . . . . .Justicia brandegeana
Shrub verbena . . . . .Lantana camara
Silk cotton . . . . .Ceiba pentandra
Silver poui . . . . .Tabebuia argentia
Sky flower . . . . .Thunbergia grandiflora
Sleepy mallow . . . . .Malaviscus arboreus
Slipper flower . . . . .Pedilanthus tithymaloides
Slipper spurge . . . . .Pedilanthus tithymaloides
Snake plant . . . . .Sansevieria trifasciata
Snakewood . . . . .Cecropia palmata
Snowbush . . . . .Breynia disticha
Sour sop . . . . .Annona muricata
Southern yew . . . . .Podocarpus macrophyllus
Spanish bayonet . . . . .Yucca aloifolia
Spanish moss . . . . .Tillandsia usneoides
Spathiphyllum . . . . .Spathiphyllum clevelandii
Spider flower . . . . .Hymenocallis littoralis
Spider lily . . . . .Hymenocallis littoralis
Split-leaf philodendron . . . . .Monstera deliciosa
Stapelia . . . . .Stapelia gigantea
Star fruit . . . . .Averrhoa carambola
Star jasmine . . . . .Jasminium multiflorum
Star jasmine . . . . .Jasminum nitidum
Starfish flower . . . . .Stapelia gigantea
Stephanotis . . . . .Stephanotis floribunda
Strangle weed . . . . .Cuscuta spp.
Strangler figs . . . . .Ficus spp.
Strelitzia . . . . .Strelitzia reginae
Striped dracena . . . . .Dracaena deremensis
Sugar apple . . . . .Annona squamosa
Sugar cane . . . . .Saccharum officinarum
Sutter's gold . . . . .Asystasia gangetica
Sweet briar . . . . .Acacia tortuosa
Sweet sop . . . . .Annona squamosa

T

Tahitian gardenia . . . . .Gardenia taitensis
Tamarind . . . . .Tamarindus indica
Tan tan . . . . .Leucaena glauca
Tangerine . . . . .Citrus reticulata
Tannia . . . . .Alocasia macrorrhiza
Tapioca . . . . .Manihot esculenta
Taro . . . . .Colocasia esculenta
Taro . . . . .Alocasia macrorrhiza
Thornbush . . . . .Acacia tortuosa
Thryallis . . . . .Galphimia glauca
Thunbergia . . . . .Thunbergia grandiflora
Ti . . . . .Cordyline terminalis
Tiare . . . . .Gardenia taitensis
Tibet . . . . .Albizia lebbeck
Tiger's claw . . . . .Erythrina variegata
Torch ginger . . . . .Nicholai elatior
Tourist tree . . . . .Bursera simaruba
Travelers palm . . . . .Ravenala madagascariensis
Travelers tree . . . . .Ravenala madagascariensis

Common name	Scientific name
Tree hibiscus	Hibiscus tiliaceus
Tropical almond	Terminalia cattapa
Trumpet tree	Cecropia peltata
Trumpet tree	Tabebuia rosea
Tube flower	Clerodendrum indicum
Tuna	Opuntia repens
Turk's cap	Malaviscus arboreus
Turks cap cactus	Melocactus intortus
Turpentine	Bursera simaruba

U

Common name	Scientific name
Unguentine cactus	Aloe barbadense
Upland cotton	Gossypium hirsutum

V

Common name	Scientific name
Vada	Ficus benghalensis
Variegated laurel	Codiaeum variegatum
Variegated philodendron	Epipremnum aureum
Virgin Island peony tree	Cochlospermum vitifolium

W

Common name	Scientific name
Walking pine	Pandanus odoratissimus
Walking pine	Pandanus utilis
Water mampoo	Pisonia subcordata
Wattles	Acacia tortuosa
Wax mallow	Malaviscus arboreus
Wedelia	Wedelia trilobata
Weeping bottlebrush	Callistemon speciosus
West Indies cedar	Tabebuia rosea
West Indies mahogany	Swietenia mahogoni
West Indies mimosa	Leucaena glauca
White angel's trumpet	Brugmansia x candida
White anthurium	Spathiphyllum clevelandii
White bird of paradise	Strelitzia nicholai
White cedar	Tabebuia pallida
White mangrove	Laguncularia racemosa
White manjack	Cordia dentata
White pepper	Piper nigrum
White poui	Tabebuia pallida
Wild plantain	Heliconia spp.
Wild tamarind	Leucaena glauca
Wood rose	Merremia tuberosa

Y

Common name	Scientific name
Yellow allamanda	Allamanda cathartica
Yellow elder	Tecoma stans
Yellow love	Cuscuta spp.
Yellow morning glory	Merremia tuberosa
Yellow oleander	Thevetia peruviana
Yellow poui	Tabebuia serratifolia
Yellow sage	Lantana camara
Yellow trumpet tree	Tecoma stans
Yellow-vein bush	Pseuderanthemum reticulatum
Yuca	Manihot esculenta
Yucca	Yucca aloifolia

INDEX TO SYNONYMS AND LOCATIONS

The abbreviation "syn." means "is a synonym for". "See" means that the plant named first is mentioned in the text under the plant named second. In either case, the second name is the desired page heading.

INDEX TO RELATIONSHIPS BY FAMILY

Family names are arranged alphabetically. All plants mentioned in this book are listed alphabetically under the family to which each belongs.

A

Family	Plants
Acanthaceae	Asystasia gangetica Crossandra infundibuliformis Graptophyllum pictum Justicia brandegeana Pseuderanthemum reticulatum Thunbergia alata Thunbergia grandiflora Thunbergia laurifolia
Agavaceae	Agave americana Agave missionum Cordyline terminalis Dracaena deremensis Sansevieria trifasciata
Amaryllidaceae	Hymenocallis littoralis
Anacardiaceae	Mangifera indica Schinus molle Schinus terebenthifolius
Annonaceae	Annona cherimola Annona muricata Annona reticulata Annona squamosa
Apocynaceae	Allamanda cathartica Carissa grandiflora Catharanthus roseus Nerium oleander Plumeria alba Plumeria obtusa Plumeria rubra Tabernaemontana divaricata Thevetia peruviana
Araceae	Alocasia macrorrhiza Anthurium andraeanum Colocasia esculenta Dieffenbachia spp. Epipremnum aureum Monstera deliciosa Spathiphyllum clevelandii Syngonium podophyllum
Araliaceae	Brassaia actinophylla Polyscias guilfoylei
Araucariaceae	Araucaria araucana Araucaria columnaris Araucaria heterophylla
Asclepiadaceae	Calotropis gigantea Calotropis procera Cryptostegia grandiflora Cryptostegia madagascariensis Stapelia gigantea Stephanotis floribunda

B

Family	Plants
Bignoniaceae	Crescentia cujete Jacaranda mimosifolia Kigelia pinnata Podraena ricasoliana Pyrostegia ignea Spathodea campanulata Tabebuia argentia Tabebuia pallida Tabebuia rosea Tabebuia serratifolia Tecoma stans Tecomaria capensis
Bixaceae	Bixa orellana
Bombacaceae	Adansonia digitata Ceiba pentandra
Boraginaceae	Cordia dentata Cordia sebestena Messerschmidia argentia
Bromeliaceae	Ananas comosus Tillandsia usneoides
Burseraceae	Bursera simaruba
Byttneriaceae	Theobroma cacao

C

Family	Plants
Cactaceae	Cephalocereus royenii Hylocereus undatus Melocactus intortus Opuntia repens Opuntia rubescens
Caricaceae	Carica papaya
Casuarinaceae	Casuarina equisetifolia
Cochlospermaceae	Cochlospermum vitifolium
Combretaceae	Conocarpus erecta Laguncularia racemosa Terminalia cattapa
Commelinaceae	Rhoeo spathacea Setcreasea pallida
Compositae	Senecio confusus Wedelia trilobata
Convolvulaceae	Ipomoea fistulosa Ipomoea horsfalliae Ipomoea pes-caprae Ipomoea quamoclit Merremia tuberosa
Crassulaceae	Kalanchoe blossfeldiana
Cuscutaceae	Cuscuta spp.
Cycadaceae	Cycas spp. Dioon spp. Encephalartos spp. Zamia spp.

E

Family	Plants
Erythroxylaceae	Erythroxylum coca
Euphorbiaceae	Acalypha hispida Acalypha wilkesiana Breynia disticha Codiaeum variegatum Euphorbia lactea Euphorbia leucocephala Euphorbia milii Euphorbia pulcherrima Euphorbia tirucalli Hevea spp. Hippomane mancinella Hura crepitans Jatropha integerrima Jatropha multifida Jatropha podagrica Manihot esculenta Pedilanthus tithymaloides Ricinus communis

G

Family	Plants
Goodeniaceae	Scaevola frutescens
Gramineae	Bambusa vulgaris Saccharum officinarum
Guttiferae	Clusia rosea Mammea americana

H

Family	Plants
Heliconiaceae	Heliconia spp.

L

Lauraceae	Cinnamonium zeylanicum
	Persea americana
Lecythidaceae	Couroupita guianensis
Leguminosae	Abrus precatorius
	Acacia tortuosa
	Albizia lebbeck
	Bauhinia purpurea
	Bauhinia variegata
	Caesalpinia coriaria
	Caesalpinia pulcherrima
	Calliandra haematocephala
	Canavalia maritima
	Cassia alata
	Cassia didymobotrya
	Cassia fistula
	Cassia grandis
	Cassia spp.
	Cassia surattensis
	Clitoria ternatea
	Delonix regia
	Erythrina crista-gallii
	Erythrina variegata
	Gliricidia sepium
	Leucaena glauca
	Parkinsonia aculeata
	Samanea saman
	Strongylodon macrobotrys
	Tamarindus indica
Liliaceae	Aloe barbadense
	Ophiopogon japonicus
	Yucca aloifolia
Loranthaceae	Phoradendron spp.
	Viscum album

M

Malpighiaceae	Galphimia glauca
	Tristellateia australasiae
Malvaceae	Gossypium hirsutum
	Hibiscus rosa-sinensis
	Hibiscus schizopetalus
	Hibiscus syriacus
	Hibiscus tiliaceus
	Malaviscus arboreus
	Thespesia populnia
Melastomataceae	Medinella magnifica
	Tibouchina urvilleana
Meliaceae	Swietenia mahogoni
Moraceae	Artocarpus altilis
	Artocarpus heterophyllus
	Cecropia palmata
	Cecropia peltata
	Ficus benghalensis
	Ficus elastica
	Ficus lyrata
	Ficus religiosa
	Ficus spp.
Moringaceae	Moringa pterygosperma
Musaceae	Musa x paradisiaca
Myristicaceae	Myristica fragrans
Myrtaceae	Callistemon speciosus
	Melaleuca quinquenervia
	Psidium guajava

N

Nyctagenaceae	Bougainvillea spectabilis
	Bougainvillea glabra
	Pisonia subcordata

O

Ochnaceae	Ochna serratula
Oleaceae	Jasminium multiflorum
	Jasminum nitidum
	Jasminum officinale
Orchidaceae	Orchids
Oxalidaceae	Averrhoa bilimbi
	Averrhoa carambola

P

Palmae	Caryota mitis
	Cocos nucifera
	Latania loddigesii
	Roystonea regia
	Veitchia merrillii
Pandanaceae	Pandanus odoratissimus
	Pandanus utilis
Passifloraceae	Passiflora spp.
Piperaceae	Piper nigrum
Plumbaginaceae	Plumbago auriculata
	Plumbago indica
Podocarpaceae	Podocarpus macrophyllus
Polygonaceae	Antigonon leptopus
	Coccoloba uvifera
	Sesuvium portulacastrum
Punicaceae	Punica granatum

R

Rhizophoraceae	Rhizophora mangle
Rubiaceae	Coffea arabica
	Gardenia taitensis
	Ixora coccinea
	Morinda citrifolia
	Mussaenda spp.
Rutaceae	Citrifortunella x mitis
	Citrus aurantifolium
	Citrus aurantium
	Citrus grandis
	Citrus limon
	Citrus paradisi
	Citrus reticulata
	Citrus spp.
	Fortunella spp.

S

Sapindaceae	Blighia sapida
Sapotaceae	Manilkara zapota
Scrophulariaceae	Russelia equisetiformis
Simaroubiaceae	Suriana maritima
Solanaceae	Brugmansia x candida
	Capsicum sp.
	Solandra maxima
Sterculiaceae	Dombeya wallichii
Strelitziaceae	Ravenala madagascariensis
	Strelitzia nicholai
	Strelitzia reginae

V

Verbenaceae	Avicennia nitida
	Clerodendrum indicum
	Clerodendrum paniculatum
	Clerodendrum thomsoniae
	Holmskoldia sanguinea
	Lantana camara
	Petrea volubilis

Z

Zingiberaceae	Alpinia purpurata
	Alpinia zerumbet
	Nicholai elatior
Zygophyllaceae	Guaiacum officinale